D1329460

Developing
Computational Skills

1978 Yearbook

Marilyn N. Suydam
1978 Yearbook Editor
Ohio State University

Robert E. Reys
General Yearbook Editor
University of Missouri

National Council of Teachers of Mathematics

Fourth printing 1987

Library of Congress Cataloging in Publication Data:

Main entry under title:

Developing computational skills.

(Yearbook—National Council of Teachers of
Mathematics; 1978)
Bibliography: p.
Includes index.
1. Arithmetic—Study and teaching—Addresses,
essays, lectures. I. Suydam, Marilyn N. II. Reys,
Robert E. III. National Council of Teachers of
Mathematics. IV. Series: National Council of
Teachers of Mathematics. Yearbook; 1978.
QA1.N3 1978 [QA135.5] 510'.7s [513'.07'1] 77-28831
ISBN 0-87353-121-3

Printed in the United States of America

Table of Contents

1. How Computational Skills Contribute to the Meaningful
   Learning of Arithmetic
   Katherine B. Hamrick, Augusta College, Augusta, Georgia
   William D. McKillip, University of Georgia, Athens, Georgia

   The changing role of computational skills is discussed. A
   level of computational skill desirable in light of predictable
   future needs is proposed, and four reasons for advocating
   that level are discussed.

2. Using Thinking Strategies to Teach the Basic Facts
   Edward C. Rathmell, University of Northern Iowa,
   Cedar Falls, Iowa

   A framework for organizing instruction on the basic facts is
   presented. Guidelines are given for blending work with concrete materials, thinking strategies, and drill into a meaningful
   program.

3. Games: Practice Activities for the Basic Facts
   Robert B. Ashlock, University of Maryland, College
   Park, Maryland
   Carolynn A. Washbon, University of Maryland, College
   Park, Maryland

   The use of games to provide practice on specified objectives
   is discussed. Four games are described in relation to several
   guidelines.

> How teachers can diagnose difficulties with computation is discussed, with emphasis on how to develop diagnostic tests. How to use such tests to plan remediation is considered in terms of several types of errors.

> A variety of student work procedures and answer patterns for whole-number computation is examined. Specific suggestions for helping students who make each type of error are offered.

> The need to include proficiency with estimation and mental arithmetic as goals for the study of computation is presented. Approaches and guidelines are described for developing these skills.

> Capturing junior high school students' interest in computation is the aim. Three activities are described in which computation acts as a needed tool within a problem-solving context.

> Ways to use a calculator as an integral part of mathematics instruction in existing curricula are provided. Activities are proposed that will help children think about mathematics rather than merely push buttons.

Preface

Topics are selected for NCTM yearbooks several years in advance of the publication date. In selecting the topic for the 1978 Yearbook, the Publications Committee showed particular foresight. The development of computational skills is of continuing concern to teachers and students, but currently it is being given renewed emphasis as public concern focuses on the achievement of certain of those skills.

Mathematics includes far more than computation: a mathematics program devoted exclusively to computation does not prepare learners for the world in which they live. But numbers and operations with numbers form the basis of many mathematics programs. In the immediate future, as in the past, the attainment of computational skills will continue to be a viable mathematical goal. The rationale for including them in the curriculum may change; for example, learning to think algorithmically, rather than merely learning individual algorithms, seems to be increasingly important.

Under the Council's new yearbook policy, a comprehensive, definitive treatment of a topic is not the goal. Yet in this yearbook readers will find articles on a broad spectrum of computational topics. The first article considers the role of computational skills in the world of today and tomorrow. Several articles focus attention on how to help children master the basic facts. Algorithms for operations with whole and fractional numbers are considered in a number of articles, with suggested instructional sequences and teaching strategies. Assessing instruction to plan for diagnosis and remediation is discussed from three points of view, with specific difficulties pinpointed. The vital need to include estimation and mental arithmetic skills as computational goals is highlighted by an article proposing both specific procedures and guidelines. Embedding computational practice in other mathematical settings is the focus of another article, and finally, ways of using a calculator as a tool in developing mathematical ideas are presented. Each article may be read independently of the others; readers are encouraged to check the Table of Contents and "dip or dive" into whatever is of greatest interest to them at a particular time.

The development of this yearbook involved literally hundreds of persons in addition to the authors, and we wish there were space to acknowledge each of them individually. Since there is not, we hope that each of those involved in each of the stages will accept our appreciation for so willingly helping to design and produce this publication. First, more than a hundred persons responded, often at length, to a request for suggestions on the scope that the yearbook should encompass. The first draft of the guidelines for proposed articles was reviewed by additional persons. At least five Council members reviewed each of the sixty outlines received. A working committee analyzed these reviews and selected the outlines that seemed most appropriate for the yearbook. The authors then completed manuscripts, each of which was read by at least three persons in addition to the working committee members. Finally, the staff in the Reston office competently handled the time-consuming tasks involved in preparing the manuscripts for publication. To each of these, we say thank you for your willingness, enthusiasm, and expertise.

One person must be cited specifically, for he was a continuous source of willing help as this yearbook was developed. We express our appreciation to Paul Trafton for his cheerful dedication at all stages. In addition, he was coauthor of an article on computational skills, published in the November 1975 *Arithmetic Teacher,* discussing ten tenets on the teaching of computation. These tenets seem sufficiently relevant to repeat in this yearbook: they may focus the reader's attention on many points made in succeeding articles.

We hope the yearbook will help teachers with some practical considerations for developing computational skills. The authors express divergent viewpoints, gleaned from their own teaching experiences and from working with other teachers. Perhaps their suggestions not only will help improve mathematics instruction in our classrooms but will encourage others to share what they have learned in journal articles and future yearbooks.

MARILYN N. SUYDAM
1978 Yearbook Editor

ROBERT E. REYS
General Yearbook Editor

Tenets

on the Teaching of Computation

1. Computational skill is one of the important, primary goals of a school mathematics program.

2. All children need proficiency in recalling basic number facts, in using standard algorithms with reasonable speed and accuracy, and in estimating results and performing mental calculations, as well as an understanding of computational procedures.

3. Computation should be recognized as just one element of a comprehensive mathematics program.

4. The study of computation should promote broad, long-range goals of learning.

5. Computation needs to be continually related to the concepts of the operations and both concepts and skills should be developed in the context of real-world applications.

6. Instruction on computational skills needs to be meaningful to the learner.

7. Drill-and-practice plays an important role in the mastery of computational skills, but strong reliance on drill-and-practice alone is not an effective approach to learning.

8. The nature of learning computational processes and skills requires purposeful, systematic, and sensitive instruction.

9. Computational skills need to be analysed carefully in terms of effective sequencing of the work and difficulties posed by different types of examples.

10. Certain practices in teaching computation need thoughtful reexamination.

How Computational Skills Contribute to the Meaningful Learning of Arithmetic

Katherine B. Hamrick
William D. McKillip

THE ROLE of computational skill in mathematics, society, and technical applications is changing. In the technical fields, such as engineering, biology, chemistry, physics, and mechanics, calculation is increasingly being done by calculators or computers. Even the slide rule, formerly the badge of the engineering student, has almost joined Napier's bones as a historical curiosity. In society, as in technology, machines are increasingly and, indeed, almost universally used in situations that require rapid and accurate computation. In everyday computational situations, such as balancing a checkbook or figuring taxes, more and more people are using a calculator.

In view of the decreasing use of hand computation in society, how can we justify a continued stress on computational skill in the elementary school? This article will indicate the level of skill that is desirable in light of predictable future needs and will present arguments that that level is mathematically and socially useful. In advocating that computational skill is still a valid objective of elementary education, we must avoid using insubstantial reasons to keep topics in our curriculum regardless of changes in society. Even though computers and calculators are rapidly changing the way most people deal with arithmetical situations, there are computational skills that continue to be educationally and socially desirable. These skills are valid objectives of elementary education.

A Desirable Level
of Computational Skill

It is impossible to describe all the computations that pupils should be able to do and to state at what speed and with what accuracy they should be done. Teachers vary in the level of skill they seek, and children of course vary in the level of skill they attain. Some examples illustrate the sort of computational skill intended.

First, let us look at the learning of the basic facts—the 390 addition, subtraction, multiplication, and division combinations. As pupils encounter these facts in the course of the arithmetic curriculum and as they develop a firm understanding of the operations involved, they should memorize these basic facts to the point of immediate, unaided recall. Consider the multiplication facts. Children "understand" multiplication when they can find products of one-digit factors using such aids as the number line, sets of objects, rectangular arrays, and repeated addition and when they can use these aids to find answers to problems in which multiplication is the appropriate operation. After they have attained this level of understanding, they should memorize the multiplication facts. A similar analysis could be made for the other operations.

Second, pupils should be able to do the standard computational algorithms for addition, subtraction, multiplication, and division with understanding and at a moderate rate of speed. They can demonstrate their understanding by explaining each step in the algorithm. For example, they should be able to reason their way through the division of a four-digit number by a two-digit number, explaining the steps and obtaining the correct answer. Children can demonstrate their understanding (1) by relating the algorithm to successive subtraction or to the partitioning of a set into equal subsets and (2) by recognizing problems for which division is the appropriate operation. It seems unnecessary, however, to practice division to the extent that they can divide an eight-digit number by a four-digit number with great speed and accuracy; certainly they can be relieved of attaining this high level of skill, which essentially turns the student into a calculator, albeit a slow and inaccurate one. However, the modest level of skill described above is still desirable, both in division and in other algorithms.

Third, skill should be developed in the areas of estimating, rounding, mental computation, and judging an answer's reasonableness. These skills are given scant attention in most classrooms because of the pressure to attain pencil-and-paper skills; yet they are necessary for using a calculator effectively. In fact, a calculator can be used to aid in their development. The level of skill in these areas should exceed that required for computation. For example, even though students may not need to attain a high degree of speed and accuracy in dividing an eight-digit number by a four-digit number, it is still desirable for them to be able to estimate the answer—for example, to know that it will be between 1 000 and 10 000. The estimate will also serve as a partial check on the exact answer obtained from a calculator if an exact answer is required.

The following quotation from volume 2 of the report of the Euclid (Ohio) Conference on Basic Mathematical Skills and Learning suggests about this same level of computational skill and also highlights the need to seek further clarification of the question about what level of skill students should attain ("Report of the Working Group . . ." 1975, p. 18):

> Students should not become completely dependent on calculators. While avoiding endless and mindless drill in computation, we should emphasize the mathematical principles and concepts underlying the computation algorithms. For example, the two-by-one digit multiplication algorithm depends on distributivity. Learning the processes of computation combined with the skills of estimation and approximation is useful in terms of readiness for future learning.
>
> We must find the "right" combination of understandings and skills to enable a student to develop an algorithm when necessary and to use the mechanical and electronic devices when it is efficient to do so. Students must know the basic single-digit number facts, including the multiplication table, and should be fluent at some relatively simple types of computation. *Exactly how much, between this "bare bones" minimum and the amount of computation that is currently being taught, is a question that needs further study.* [italics added]

The level of computational skill described above is modest in comparison with the level some teachers attempt to obtain and is also substantially below what one would infer from an examination of school textbooks. What are the reasons for advocating the attainment of this level of computational skill? There are at least four. First, it facilitates the learning of subsequent related topics. Second, computational skill helps pupils to understand both the meaning and the significance of arithmetic operations and to apply these operations appropriately. Third, it facilitates an exploration of various topics, generalizations from data, and the recognition of generalizations. Fourth, some aspects of computational skill continue to have considerable social utility. These four reasons will be explained and some examples presented for each.

Computational Skill Facilitates Meaningful Learning of Both Concepts and More Advanced Skills

One common presentation of multiplication is to define it as "the repeated addition of the same number" or as "a short form of addition." This is only one of several meanings that are a part of the concept of multiplication, but it is an important meaning. It is universally taught, and at a *verbal level* it is almost universally learned. Fordham (1974) has found that in one school virtually every fourth grader could state this relation between addition and multiplication, although many could not use it.

Of course, if Larry, for example, does not have sufficient computational skill in addition, he cannot use the "repeated addition" meaning of 5 × 7 (i.e., 7 + 7 + 7 + 7 + 7) even though he may know it. He is effectively prevented by his lack of computational skill in addition from using meaningful ways of recovering a forgotten multiplication fact. Larry could not, for example, use the addition of five sevens, as above, nor could he use a known multiplication fact, 4 × 7 = 28, to find the answer to 5 × 7 (i.e., 28 + 7).

Let us examine his all-too-typical work on a multiplication problem (fig. 1.1). The five rows of seven marks make it evident that he understands one meaning of multiplication. For Larry, addition *is* counting— that is, he still counts to find answers to basic addition facts. Multiplication will become an extended counting process, and the multiplication algorithm will become extremely tedious. Immediate attention to learning basic facts, first the addition facts and then those of the other operations as they are presented, would provide this pupil with the basis needed to make use of the relation between addition and multiplication in appropriate situations.

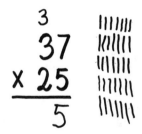

Fig. 1.1

In some instances the meaning of a computational procedure and the way it produces an answer can be established through hand computation. Then the algorithm can be continued by using a calculator when the calculation becomes long or tedious. An example of this is the procedure for finding a square root (fig. 1.2).

Find the square root of 56. (A reasonable first estimate is 7.)

$$
\begin{array}{r}
8 \\
\hline
7\,)\overline{56}
\end{array}
\qquad
\begin{array}{r}
8 \\
+7 \\
\hline
15
\end{array}
\qquad
\begin{array}{r}
7.5 \\
\hline
2\,)\overline{15.0}
\end{array}
$$

$$
\begin{array}{r}
7.46 \\
\hline
7.5\,)\,\overline{56.0\ 00} \\
52\,5 \\
\hline
3\,5\,0 \\
3\,0\,0 \\
\hline
5\,0\,0 \\
4\,5\,0 \\
\hline
\end{array}
\qquad
\begin{array}{r}
7.46 \\
+7.5 \\
\hline
14.96
\end{array}
\qquad
\begin{array}{r}
7.48 \\
\hline
2\,)\overline{14.96}
\end{array}
$$

Fig. 1.2

The remarkable speed with which this algorithm converges on the square root can be seen in the result thus far; an estimate of 7.48 cannot be more than 20 hundredths off. At this point we declined to do further hand computation but took another step using a calculator (see fig. 1.3). The calculator is lavish with decimal places, and there is no need to round off until the end of the calculation, when 7.48331 is obtained. Understanding this algorithm depends on a knowledge of the nature of the number sought—the square root—and on a knowledge of the relation between multiplication and division.

$56 \div 7.48 = 7.486631$ $\qquad$ $7.486631 + 7.48 = 14.966631$

$14.966631 \div 2 = 7.4833155$ $\qquad$ $56 \div 7.4833155 = 7.483314$

Fig. 1.3

To complete this analysis of the relation between hand and machine calculation, it would be valuable to have pupils write a flowchart for this iterative calculation and follow the steps of the flowchart using the calculator. Where a programmable machine is available, they could program the machine to do the calculation.

As these examples show, *computational skill in one operation can contribute to a pupil's understanding of a more advanced operation and to the understanding of concepts.*

Not enough just to understand concept. must be able to do it (comp. skill)
understand + (mem, basic facts)

Computational Skill Helps Pupils Understand Arithmetic Operations and Their Applications

A child who understands the meaning of an operation can relate the operation to an action on sets or to another operation. For example, "4 × 5" can be related to the action of combining four sets of five objects or to summing 5 + 5 + 5 + 5. The child who does not have recall of addition facts or computational skill in addition will be unable to utilize the addition meaning of multiplication. The pupil will view finding 4 × 5 as a tedious counting process.

It has been noted that computational skill with one operation facilitates the learning of the meaning of other operations. Computational skill also helps students apply arithmetic operations appropriately in problem situations. For example, what is the appropriate operation to use to answer the question "How many Xs?" in figure 1.4? There are three common answers: multiplication (4 × 6 or 6 × 4), addition (6 + 6 + 6 + 6 or 4 + 4 + 4 + 4 + 4 + 4), or counting. We expect our pupils to look at this example as a straightforward application of multiplication. However, for the child who does not know multiplication facts, it is *not* a multiplication problem. The child who lacks multiplication skill but has computational skill in addition would see it as an addition problem. For those who lack computational skill in both multiplication and addition, this example is only a counting exercise.

```
    X       X       X       X       X       X

    X       X       X       X       X       X

    X       X       X       X       X       X

    X       X       X       X       X       X
```

Fig. 1.4. How many Xs?

Of course, knowing the meaning of the operation of multiplication also plays an important role in the ability to apply multiplication appropriately. However, in the study by Fordham (1974), several pupils demonstrated a knowledge of the meaning of multiplication by relating a multiplication number sentence to repeated addition, an array, or other model but could not apply multiplication appropriately. These pupils lacked recall of multiplication facts. But all the children in the study who knew the meaning of multiplication *and also had good recall of multiplication facts* were able to apply multiplication appropriately. Children's computational skill as well as their knowledge of the meaning of each operation enables them to apply the operation appropriately.

A second example of an application of multiplication typical of many textbooks is the following problem:

> Mrs. Jones bought 5 cartons of cola. There were 6 bottles in each carton. How many bottles did she buy?

Although we are trying to provide practice in applying multiplication, Jane, who lacks computational skill in multiplication, might do this as the addition problem 6 + 6 + 6 + 6 + 6 and obtain the correct answer; she sees this "application of multiplication" problem as an addition problem. The objective of having our pupils work application problems is not only to have them find answers but also to have them *recognize each problem as an application of a particular operation.* In order to accomplish this, we must first ensure that they have the computational skill to actually use the operation to find the answer to the problem.

Computational Skill Facilitates the Exploration of Topics, Generalizations from Data, and the Recognition of Generalizations

For these purposes computational skill is in many ways superior to a calculator. Lessons in which pupils use their computational skill to construct examples and then generalize from these examples provide for more pupil participation and involvement than those presented by the teacher. For example, one way to begin developing concepts in number theory is through a list of numbers and their factors. It is clear that generating the list in table 1.1 depends both on an understanding of the concept of factor and on computational skill in multiplication.

TABLE 1.1
NUMBERS AND THEIR FACTORS

1:	1	10:	1, 2, 5, 10
2:	1, 2	11:	1, 11
3:	1, 3	12:	1, 2, 3, 4, 6, 12
4:	1, 2, 4	13:	1, 13
5:	1, 5	14:	1, 2, 7, 14
6:	1, 2, 3, 6	15:	1, 3, 5, 15
7:	1, 7	16:	1, 2, 4, 8, 16
8:	1, 2, 4, 8	17:	1, 17
9:	1, 3, 9	18:	1, 2, 3, 6, 9, 18

We could, of course, produce the list for our nonmultiplying students, but this would make the lesson "ours"; the pupils would have little part in it except as observers. It is much better for them to exercise their skill in multiplication to produce the list. A calculator could be used, but unless the pupils already have skill in factoring (thus using the calculator as only a checking device), the procedure would be very tedious.

By examining table 1.1, pupils can note several basic ideas: In the list there is one number having only one factor. Could there ever be another? Why or why not? There are some numbers having exactly two factors. List them. These are called "prime numbers." Can you find some more? Is 101 a prime? 102? Is 1 a prime number? How is 1 different from the primes?

Suppose we ask, "What numbers have an odd number of factors?" The list will include 1 (one factor), 4 (three factors), 9 (three factors), and 16 (five factors). What is it that these numbers have in common? To our practiced eyes it is "obvious" that each of these numbers is the product of two equal whole-number factors. We identify this as a list of the first four perfect squares, and the list could be extended. We hypothesize that the numbers with an odd number of factors are perfect squares. However, without a knowledge of the multiplication facts it is extremely difficult to see the common property and then extend the list so that the hypothesis can be tested.

These explorations, formulating and testing hypotheses and *participating in the lesson* rather than merely observing, are dependent on skill in multiplication.

Suppose we want pupils to realize that whole numbers can be factored, as products of primes, in one and only one way. The children first use their knowledge of multiplication in factoring to construct factor trees. For example, the composite number 60 could be factored into 6 × 10 (fig. 1.5). The factors 6 and 10 can be factored into 2 × 3 and 2 × 5 respectively. Since 2, 3, and 5 are prime numbers, the factor tree is complete.

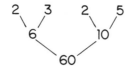

Fig. 1.5. Factor tree for 60

If a group of children constructed factor trees for the composite number 60, several different factor trees would probably be developed (fig. 1.6).

The numbers at the ends of the branches in figure 1.6 are the same—2, 2, 3, and 5—for all trees. Only the order of the numbers differs from

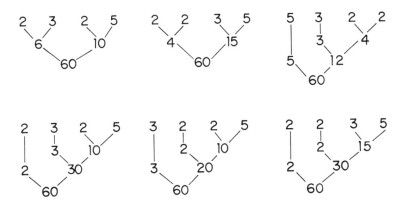

Fig. 1.6.   Factor trees for 60

tree to tree. The formation of factor trees for other composite numbers and guiding questions from the teacher could lead the children to such generalizations as "No matter how you factor a number, you get the same list of prime factors." A conclusion of this type would imply an understanding of the fundamental theorem of arithmetic: *Except for the ordering of the primes, a composite number can be expressed as a product of primes in only one way.* A knowledge of multiplication facts permits the pupils to participate in the lesson. Without such skill, they are only observers of the lesson.

Computational skill in multiplication also helps pupils to make generalizations in geometry. One way to introduce the concept of volume is to look at "boxes" similar to the ones in figure 1.7.

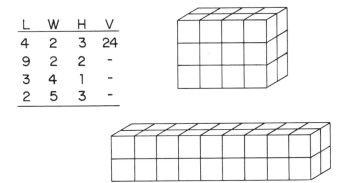

L	W	H	V
4	2	3	24
9	2	2	-
3	4	1	-
2	5	3	-

Fig. 1.7.   Boxes made with one-centimeter cubes

Different boxes of specified length, width, and height are actually constructed by the children with unit cubes. Volume can then be defined simply as the number of one-centimeter cubes (cubic centimeters) used to construct the box. After completing the chart in figure 1.7, those pupils having computational skill in multiplication can "see" the volume as a product of length, width, and height.

In some problem situations the frustration produced by tedious calculation stimulates mathematical discovery. The classical problem

$$1 + 2 + 3 + 4 + \ldots + 98 + 99 + 100 = ?$$

can be done on a calculator in a matter of moments. If students are to be led to exercise their mathematical inventiveness, however, calculators must be *un*available when this problem is presented. After the students have been led to a clever solution such as adding the series twice (fig. 1.8) and have applied the method to other such problems, the results can be verified with calculators. The method is so neatly plausible, however, that if the calculator does not verify the result, it should be suspected that a mistake was made in using the calculator!

$$
\begin{array}{c}
1 + 2 + 3 + 4 + \ldots + 98 + 99 + 100 \\
100 + 99 + 98 + 97 + \ldots + 3 + 2 + 1 \\
\hline
101 + 101 + 101 + 101 + \ldots + 101 + 101 + 101 = 100 \times 101
\end{array}
$$

$$\text{hence, } 1 + 2 + 3 + \ldots + 98 + 99 + 100 = \frac{100 \times 101}{2}$$

Fig. 1.8

Thus in a variety of exploratory activities and in some problem-solving activities the exercise of some computational skill makes it possible for a student to seek a generalization and to recognize a generalization in numerical information.

Computational Skill
Still Has Considerable Social Utility

It is true that with the increased use of calculators, large numbers of people get along without much computational skill. However, this does not reduce the force of the argument that some computational skill is very handy. There are some situations in which these skills are more suited to the need at hand than a calculator is.

The main objective in most of these social situations is to produce a ball-park estimate of some result. This result may or may not be worked out exactly later. Estimation is useful in making decisions and in determining whether or not it is important to get an exact answer. Number sense, and in particular the skill of rounding to obtain an approximation, is the first step in many calculations. Quite frequently a knowledge of basic facts is needed to complete the calculation, and the check is to ask whether the answer is reasonable under the circumstances.

How long will it take us to get there? is a question asked frequently but not requiring an exact answer. Indeed, no exact answer is possible because of many unpredictable circumstances. In figure 1.9 the "story problem" approach is contrasted with the sort of calculation adults might actually do under the circumstances. Even though they have not done a written calculation, adults recognize that the situation calls for division, understand the connection between $35 \div 5$ and $350 \div 50$, and know the basic division fact. If the answer had come out 2 hours or 20 hours, it would have been rejected as unreasonable under the circumstances.

John drives 335 miles at 50 miles an hour.
How long does it take to drive that distance?

$$
\begin{array}{r}
6.7 \\
50 \overline{)335.0} \\
\underline{300} \\
350 \\
\underline{350}
\end{array}
$$

6.7 hours = 6 hours 42 min.

We need to drive a little under 350 miles; we will average about 50.

$$35 \div 5 = 7$$
So $350 \div 50 = 7$

The trip will take a little under 7 hours.

Fig. 1.9

Most people who shop for groceries keep some track of the cost as they shop. The techniques for doing this are quite varied, but rounding to the nearest dime and adding are common. Grouping items in approximately half-dollars and dollars is another common technique. When one examines these useful manipulations (fig. 1.10), it is apparent that an understanding of quantity and the ability to perform simple additions mentally are sufficient to give a reasonably close estimate of the cost.

Nearest dime		Dollars and half dollars
Think:	$ .79	Think:
$1.20	.35	Just over $1.00
$2.60	1.42	about $2.50
$3.30	.69	over $3.00
$4.80	1.51	over $4.50
$5.80	.99	over $5.50
$6.00	.17	near $6.00
$6.50	.46	about $6.50
	$6.38	

Fig. 1.10

Although little attention has been given to exact hand computation, there are situations in which this skill is useful. Calculating one's part of a group restaurant bill is a common situation. A few purchases at a drugstore may still be totaled by hand. These computations are not to be ignored. However, the increased emphasis on rounding, estimation, and the reasonableness of results, long advocated, is a pressing issue in the elementary school curriculum because there are many circumstances where these skills are valuable.

Summary

The development of computational skill is a desirable objective for mathematics in the elementary grades. It facilitates the teaching of many topics; it helps children to understand both the meaning and the significance of the mathematics they are learning; it facilitates the exploration of topics and the recognition of generalizations; it has social utility.

If most pupils are to achieve and use this level of skill, a shift in emphasis will be necessary. More attention must be paid to understanding the meaning of the operations and to understanding why an algorithm works to produce a reasonable answer. Less attention may be given to performing long computations with great speed. Much more attention must be given to learning the basic facts, performing mental computation, and estimating results.

REFERENCES

Fordham, D. L. "An Investigation of Third, Fourth, and Fifth Graders' Knowledge of the Meanings of Selected Symbols Associated with Multiplication of Whole Numbers." Doctoral dissertation, University of Georgia, 1974. (University Microfilms no. 75-8138)

"Report of the Working Group on Goals for Basic Mathematical Skills and Learning." In *Working Group Reports, Conference on Basic Mathematical Skills and Learning,* vol. 2, pp. 16–21. Washington, D.C.: National Institute of Education, 1975.

2

Using Thinking Strategies to Teach the Basic Facts

Edward C. Rathmell

MOST teachers have been confronted with children who use their fingers to figure out the basic facts. Many feel that this procedure should be allowed initially but that eventually children should *know* the basic facts for an operation. What experiences help them develop from finger counting to immediate recall of the facts?

One common attempt to solve this instructional problem is to rely heavily on drill. Although interesting games and activities might be used in drill, the long-term results of learning often do not meet teachers' expectations. Another teaching strategy involves the use of concrete materials. Children generally enjoy using the materials and soon learn to manipulate them to determine answers. But again, immediate recall of the basic facts is often not the outcome.

Why is drill effective in some situations and ineffective in others? Why are experiences with concrete materials helpful at times and seemingly unhelpful at other times? A growing body of evidence seems to indicate that the thinking strategies used by children to figure out answers may be the key.

A variety of thinking strategies can be used to solve any given fact problem. Some of these strategies involve using concrete materials, counting on fingers, or counting by ones. Others are more mature strategies in the sense that a known fact is used to figure out an unknown fact.

Besides being classified as mature or immature, thinking strategies can also be classified as efficient or inefficient. For any given fact problem, some strategies can be used efficiently to solve the problem and others cannot. The efficient strategies—efficient in the sense of the amount of time and cognitive processing required to determine the solution—vary from problem to problem. For example, simple counting strategies are just as quick and easy to use for addition facts with sums up to 5 as a more mature strategy might be. However, counting from one is not an efficient way to solve addition facts with sums greater than 10.

A conceptual framework for organizing instruction on the basic facts is presented in this article. This framework provides guidelines for blending concrete materials, thinking strategies, and drill into a meaningful and successful program. It provides a basis for deciding when activities with concrete materials can be expected to be most effective and when drill would be.

A Rationale for Teaching Thinking Strategies

In several of his writings, Brownell stated that children develop an understanding of the basic facts in a series of stages that are characterized by the thinking they use to determine answers to fact problems. Both Brownell and Jerman have conducted research that involved categorizing students accordingly. Brownell found that a child's strategy varies from problem to problem, although patterns do seem to emerge for similar problem types. Jerman's work (1970) supports this finding. Jerman also concluded that "children who have learned to use a certain strategy for a combination continue to use the same strategy as they grow older and just do it more rapidly" (p. 106).

Why do children continue to use the same strategy for a basic fact over relatively long periods of time, particularly if the strategy is inefficient? Brownell and Chazal (1935) concluded that drill on basic facts increased pupils' speed and accuracy but did not change the thinking they used to solve fact problems. Pupils who used an inefficient counting strategy before drill used the same strategy after drill, but the counting itself was performed more quickly. Some discover new thinking strategies on their own, but drill is not sufficient for most of them to develop mature and efficient methods for solving basic facts. They need explicit instruction to learn these new strategies.

How effective is it to teach children new thinking strategies? Both Thiele (1938) and Swenson (1949) found convincing evidence that teaching children different thinking strategies facilitates their learning and retention of the basic addition facts and transfer to other problems. Thornton found similar evidence for each of the four fundamental operations.[1] Thiele also noted that "the superior achievement of the pupils taught by the generalization method is largely due to the differences in achievements with the so-called harder addition facts" (1938, p. 70). According to Thornton, students who had been taught new thinking strategies had greater achievement on the harder facts. Recent data lend support to the hypothesis that children with immature thinking strategies do not perform as well on the harder addition facts as those who have developed more mature strategies. First- and second-grade pupils who were capable of using and explaining (1) how to count on and (2) how to solve fact problems that are one more or one less than a known fact were compared to those students who were unable to use or verbalize the strategies. For addition facts with sums greater than 10, the first-grade students with more mature strategies correctly answered more than twice as many problems and were nearly twice as accurate as the other first graders on a timed test. Similar second-grade differences existed but were not so extreme. At both grade levels the groups were barely distinguishable in speed or accuracy for facts with sums less than or equal to 10.

It was also found that differences existed on retention scores over the summer vacation. Students who used more mature strategies tended to retain their knowledge of the basic facts better than the others. At the second-grade level, for addition facts with sums greater than 5, the group with more mature strategies showed little or no drop on the number of correct responses on timed tests given just prior to and just after summer vacation. The other group could correctly answer only about half as

1. Carol Thornton 1977: personal communication

many items on posttests. Similar results were obtained at the first-grade level for sums from 5 to 10. For greater sums the first graders without mature strategies had such low scores on the pretest that they could show very little drop over the summer.

Teaching mature thinking strategies, as indicated in the studies by Thiele, Swenson, and Thornton, provides pupils with a more efficient attack for the harder facts. With practice, the methods can be applied easily and quickly. These strategies also enable pupils to organize and understand relations among the facts that aid in memorization and recall.

Three Components of Instruction for the Basic Facts

Children can profit from three types of experiences to help them learn the basic facts for an operation. They include (1) activities with concrete materials, (2) instruction on new thinking strategies, and (3) drill. A discussion of the role that each of these plays in the learning program is necessary in order to describe more clearly how they can be blended into a meaningful instructional program.

The role of concrete materials

Concrete materials should probably be used during the initial instruction for an operation. The main purpose of instruction at this time is not so much to teach children how to derive answers as it is to develop or broaden their concept for the operation and to develop appropriate language and symbolism.

Concrete materials play an important role in concept development. The materials become a referent for work involving the operation. They provide a link to connect the operation to real-world problem-solving situations. Materials are also used as "proof" for other work. Whenever a child is unsure of mental or symbolic work, wants to confirm that a thinking strategy really works, or simply wants to be sure that an answer is correct, materials can be used for confirmation.

Other important objectives of the experiences with concrete models during this initial stage are the development of appropriate language and symbolism. The language is learned informally as children communicate about what they are doing and what they see happening. As they use models, they should begin to understand the symbolism related to the operation. At this stage, the manipulation of symbols is not the primary objective. Instead, it is important to build a referent for each symbol.

After this initial instructional period with materials, these models can most profitably be used to help children learn new thinking strategies.

These strategies can be learned by using the materials in ways that involve new thinking.

The role of thinking strategies

The primary objective of teaching thinking strategies is to help children learn the mature strategies that are useful and efficient to solve the harder facts. This thinking enables them eventually to discard concrete materials and to begin working at the symbolic level of representation.

Thinking strategies may also be an important factor in helping children recall facts from their memory. The most important thing to do to make information retrievable is to organize it and have learners develop many relations among the facts. By learning a new thinking strategy, a child is learning new relations among the different facts. This tends to facilitate organization and to make the whole set of facts more coherent.

The role of drill activities

Drill has long been recognized as an essential component of instruction in the basic facts. Practice is necessary to develop immediate recall. Brownell and Chazal (1935) have shown quite convincingly that drill increases the speed and accuracy of responses to basic-fact problems. Those are the purposes for which drill should be used. Drill alone will not change the thinking that a child uses; it will only tend to speed up the thinking that is already being used.

Since little can be gained by speeding up immature or inefficient thinking, it seems reasonable to state that drill will be most effective when the thinking that is being used is efficient. Consequently, drill on the harder facts is more likely to be effective after children have developed appropriate thinking strategies.

This has implications for the organization of drill. On the one hand, those facts for which even immature strategies are efficient can be drilled soon after the pupils are familiar with the concrete models. On the other hand, drill on harder facts should be postponed until children have developed the efficient thinking strategies necessary to handle them. Otherwise, it may be ineffective.

Conceptual Framework

Experiences with concrete materials facilitate the development of initial concepts, encourage the use of appropriate language and symbolism, and provide a means of learning mature thinking strategies. Thinking strategies help children derive answers to fact problems and provide structure for organizing the facts so that recall will be easier. Drill helps pupils increase speed and accuracy. Together these three types of experiences can be blended into a meaningful instructional program for learning the basic facts for an operation:

Step 1.   Students should develop confidence and competence in using a model for the operation.

Step 2.   Drill can begin for the easy facts, those for which even immature thinking strategies are relatively efficient.

Step 3.   A model can be used to help pupils learn a more mature thinking strategy, one that will enable them to solve more difficult fact problems.

Step 4.   Drill can begin for those facts that can be easily solved by the thinking strategy just learned.

Step 5.   Steps 1, 3, and 4 can be repeated to introduce other models and thinking strategies as appropriate for the operation.

Step 6.   Drill can be used to develop speed and accuracy in recalling the facts.

This framework has three features that should be noted. First, within the context of this framework children will be given every opportunity to develop a strategy before being drilled on facts for which they have no efficient or quick way to figure out solutions. Situations that call for immediate response are very frustrating to children who have no thought processes available that permit them to answer in the time allotted. Second, models are introduced one at a time. This enables the pupils to gain operational comprehension with one model before having to cope with another. Third, new models and new thinking strategies are introduced in a manner that interweaves concrete and symbolic experiences. Consequently, the children are provided several opportunities to integrate the understandings and skills they have acquired. This, of course, aids in the memorization, recall, and application of the basic facts.

Applying the Framework

When instruction for the basic facts is being planned, the following key questions need to be answered:

1. What models can children best use to represent the operation?
2. What thinking strategies can they most easily use to help them learn the facts?
3. Which facts can most easily be solved by each strategy?

4. How can the models be used to introduce each thinking strategy?

5. What sequence of steps is most likely to ensure meaningful learning?

Addition

Models for addition

Two types of models are commonly used to represent addition. They are the "join sets" model and the measurement models, such as the number line or rods. Together they permit children to deal with a wide variety of addition situations, including either discrete or continuous quantities. The illustrations presented here include only examples of joining sets and the number line (fig. 2.1).

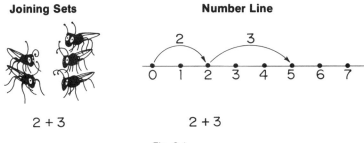

Fig. 2.1

Thinking strategies for addition

Children use a variety of thinking strategies to help them solve addition facts. Primary-grade children are commonly found to (1) *count on,* (2) recognize an unknown fact to be *one more* or *one less* than a known fact, and (3) use a sharing, or *compensation,* technique to change an unknown fact into a known fact. Some students also count their fingers or other objects. However, these three strategies seem to be the most likely ones to facilitate learning of harder facts.

One More or One Less Than a Known Fact

$5 + 6 =$ ___
$5 + 5 = 10$, SO
$5 + 6$ IS 1 MORE, OR 11.

$9 + 4 =$ ___
$10 + 4 = 14$, SO
$9 + 4$ IS 1 LESS, OR 13.

Compensation to Make a Known Fact ("Add to Ten, Then On")

$9 + 6 =$ ___
9 AND 1 IS 10, 10 AND 5 MORE
IS 15, SO $9 + 6 = 15$.

The sequence outlined here focuses on these three strategies because (1) children use them, (2) together they provide a means for easily solving nearly all the facts, and (3) they can be taught.

Facts to be learned by each strategy

Counting on is a strategy that is most efficiently used when one of the addends is small. Most children can learn to count one, two, or three more than a given number without overloading their memory. To count on more than three is difficult because it is hard for children to remember both what comes next and how many have been counted. Children should usually start with the larger addend, regardless of its position in

the sentence, and then count on from there. Counting on can be efficiently used with the shaded facts given in figure 2.2.

Recognizing that an unknown fact is *one more* or *one less* than a known
fact is commonly used for many of the harder facts. Since children are
generally able to learn the doubles and how to add 10 without any special
strategies, these facts are most often the known facts. As soon as children
have these two prerequisites, they can begin using the one-more or one-
less strategies to learn the near-doubles (consecutive numbers) and
adding 9, as shown in figure 2.3.

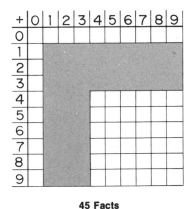

45 Facts

Fig. 2.2. Counting on

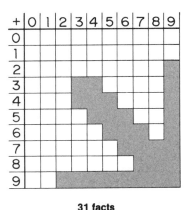

31 facts

Fig. 2.3. One More or One Less

Using a sharing, or *compensation,* technique to change an unknown
fact into a known fact also centers on 10. Both addends are changed, one
increased and the other decreased, to make one of the addends 10.
Sometimes students think, "Add to ten, then add on." For example,
9 + 6 is "changed" to 10 + 5 by thinking "9 and 1 more is 10; 10 and 5 is
15." The facts learned through compensation are shown in figure 2.4.

The facts that can be solved efficiently by using the counting-on and
the one-more or one-less strategies include all but a few of the 100 basic
facts. They are shown in figure 2.5.

When those facts that can easily be solved by using the compensation
("add to ten") strategy are also included, only the zero facts and six
others remain.

Teaching the thinking strategies

Helping children learn to use new thinking strategies takes time. They
may understand a strategy when someone else demonstrates it, but they
will need frequent practice before they are able to use it spontaneously as

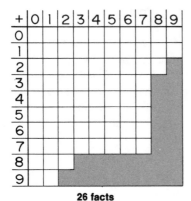

26 facts

Fig. 2.4. Compensation, or "Add to Ten"

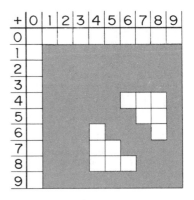

69 facts

Fig. 2.5. Counting On and One More or One Less

a method of solution. One technique that has been found effective is to (1) teach a lesson on the strategy and (2) maintain exposure to the same thinking for about two weeks by "talking through" a few examples each day, perhaps for only two or three minutes. Nearly all students who have the prerequisites for learning the strategy will be spontaneously using it by then. This time spent on helping children practice using the thinking strategies is repaid at the memorization stage of learning.

Short activities can be used to maintain exposure to new thinking strategies and help children practice using them. In order to use counting on, students must know the counting sequence so well that on being given any number less than 10, they can immediately say the next number in the sequence. Practice in quickly telling the next number is quite helpful for some children.

Since the pupils have competence in counting from one as a method of solution, it is sometimes difficult to get them to use counting on. One effective method is to cover one of two sets being joined, remind them of the number in the covered set, and have them determine how many in all. With some prompting initially, the students soon find that they are able to figure out the correct answer by counting on. They gain confidence and find that counting on is much quicker than their previous counting strategies.

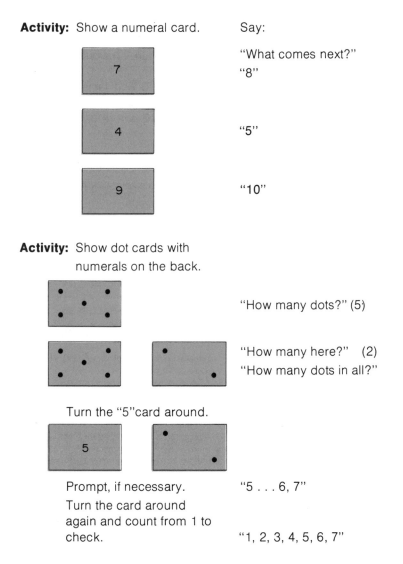

Activity: Show a numeral card.   Say:

7

"What comes next?"
"8"

4

"5"

9

"10"

Activity: Show dot cards with
                    numerals on the back.

"How many dots?" (5)

"How many here?"   (2)
"How many dots in all?"

Turn the "5"card around.

5

"Prompt, if necessary.   "5 . . . 6, 7"
Turn the card around
again and count from 1 to
check.   "1, 2, 3, 4, 5, 6, 7"

Activity: Put six marbles in your   "I have six marbles."
          pocket.

          Put one more in your   "Now how many do I have?"
          pocket.

          Prompt, if necessary.   "6 . . . 7"

          Count from 1 to check.   "1, 2, 3, 4, 5, 6, 7"

Before students can use the one-more and one-less strategies with doubles, they must know some of the doubles, usually through 6 + 6. Eventually, as they learn the doubles of 7, 8, and 9, they learn to use the strategies with those facts as well. Practice with the doubles is good preparation.

This strategy can also be used with 10 to help with facts where 9 is an addend. This depends on the students' being able to add a single-digit number to 10. Have them practice adding 10 before using the one-less strategy to help with the facts of 9.

Activity: Show two groups of 5.   "How much is 5 + 5?"

          Place 1 more with one of   "Now there are 5 and 6. How many
          the groups.   in all?"

          Prompt, if necessary.   "1 more than 10."

          Check it.   "5 + 5 = 10; so 5 + 6 is 1 more, or
                                            11."

Activity: Show a group of 6 and a   "How much is 6 + 7?"
          group of 7.

          Separate 1 from the group   "6 + 6 = 12; so 6 + 7 is 1 more, or
          of 7.   13."

          Check it.

Activity: Show two groups of 5.   "How much is 5 + 5?"

          Remove 1.   "Now there are 4 and 5. How many
                                            in all?"

          Prompt, if necessary.   "1 less than 10."

          Check it.   "5 + 5 = 10; so 4 + 5 is 1 less, or
                                            9."

Activity: Show a group of 9 and   "How much is 9 + 5?"
a group of 5.
Place 1 more with the   "10 + 5 is 15; so 9 + 5 is 1 less, or
group of 9.   14."
Check it.

The compensation, or "add to ten," strategy is more difficult for children to learn. Since both addends are changed, the memory load is significantly greater. It requires some maturity to keep track of all the changes in the addends. Nevertheless, with careful instruction, most second-grade pupils are able to learn this useful strategy. The children also need to be able to add 10 and a single-digit number quickly. Have them practice this just prior to instruction for the strategy.

Activity Sequence (developed after a discussion with William Swart):

1. Show 10 + 7 = _____   "Answer both. Which is the easy
       9 + 8 = _____   one?"
   Show 10 + 5 = _____   "Answer both. Which is the easy
       9 + 6 = _____   one?"
   Repeat and discuss   "Adding 10 is easy."

2. Ring the easy problem. Do not write the sum.

   9 + 7

   9 + 5   10 + 6   8 + 5

3. Show a group of 9 and a
   group of 4.

            "How can we change this to an easy
                                problem?"

   Make 10 by moving a block
   from the 4.

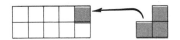   "Is this an easy problem?"

4. Change these to easy problems. Write the sum.

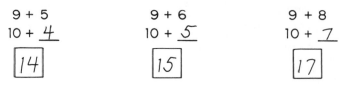

9 + 5   9 + 6   9 + 8

10 + 4   10 + 5   10 + 7

| 14 | | 15 | | 17 |

Activity: Show the number line.

Say:

"What is 8 + 4?"

"8 and 2 more is 10. Then 2 more is 12."

Check it.

An instructional plan

The following steps are one example of how the framework can be applied to ensure meaningful learning. Children are not asked to give quick responses to facts until they have been given the opportunity to develop thinking strategies that enable them to do so.

Step 1. Introduce addition by joining sets.

Step 2. Use games and other practice for sums to 5.

Step 3. Use the model to develop the strategy of counting on.

Step 4. Use games and other practice for facts with at least one small addend.

Step 5. Use the model to help children learn the doubles and adding 10, then to develop the one-more and one-less strategies, one at a time.

Step 6. Practice the doubles, the near-doubles, and the facts having 9 as one of the addends.

Step 7. Introduce the number line and review the strategies learned so far.

Step 8. Use a model to develop the add-to-ten strategy.

Step 9. Drill the remaining facts to develop fast and accurate recall.

The sequence outlined above would normally be used over a two- or three-year period in the primary grades. Each step can be roughly equated to a short unit of instruction.

Older pupils who need remedial experience with addition facts can also profit from the sequence. However, their placement should depend on the thinking strategies they use, not on the facts they know. Pupils who count on their fingers can begin by learning to count on. Pupils who count on can learn to use the one-more and one-less strategies.

Multiplication

Models for multiplication

Many different models can be used to represent multiplication. Only three are presented here: (1) the set model, that is, joining equivalent sets; (2) the array; and (3) the number line. Together these three permit children to develop a reasonably comprehensive concept of multiplication and the mature thinking strategies needed to facilitate the learning of the multiplication facts.

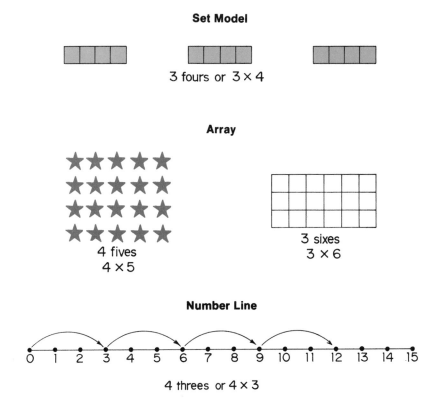

Set Model

3 fours or 3 × 4

Array

4 fives
4 × 5

3 sixes
3 × 6

Number Line

4 threes or 4 × 3

Thinking strategies for multiplication

Unlike addition, there are many strategies that students can and do use to determine answers to multiplication facts. Those illustrated here represent strategies commonly used. They provide enough diversity to facilitate the learning of any multiplication fact.

Skip Counting

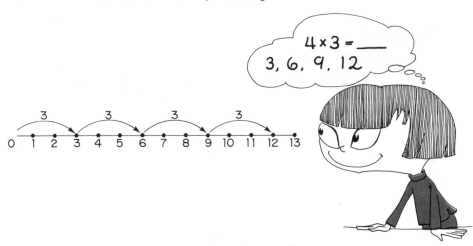

Repeated Addition

Splitting the Product into Known Parts

One More Set

Twice as Much as a Known Fact

Facts of 5

5 eights

2 eights

7 × 8 = _____
5 EIGHTS IS 40.
2 EIGHTS IS 16.
SO 7 EIGHTS IS
40 + 16, OR 56

Patterns

1	2	3	4	5	6	7	8	9	10
11	12	13	14	15	16	17	18	19	20
21	22	23	24	25	26	27	28	29	30
31	32	33	34	35	36	37	38	39	40
41	42	43	44	45	46	47	48	49	50

6 × 9 = _____
5 NINES IS 45. 6 NINES
HAS 1 MORE TEN AND 1
LESS ONE, OR 54.

6 × 9 = _____
THE SUM OF THE DIGITS IS NINE.
6 NINES IS IN THE FIFTIES, AND
5 + 4 IS 9; SO IT IS 54.

Facts to be learned by each strategy

Each strategy illustrated above can be used efficiently with some of the multiplication facts, but not all the strategies can be used efficiently with all the facts. *Skip counting* works quite well with small factors, that is, for multiples of 2 and 5, possibly for multiples of 3 and 4. However, counting by sevens is quite difficult for most pupils. *Repeated addition* can also be used efficiently when one of the factors is less than 5. However, children should not be encouraged to solve facts like 9 × 7 by repeated addition; it takes too long, and there is too great a chance for error.

Neither of the strategies above works particularly well with factors greater than 5. Both still play an important role in helping children learn multiplication facts. Children are usually competent at both counting and addition prior to the time they are introduced to multiplication. By using these strategies during the initial work with multiplication, they can make use of what they already know and get a confident start in learning this new operation.

The facts for which skip counting and repeated addition can be used efficiently are indicated in figures 2.6 and 2.7.

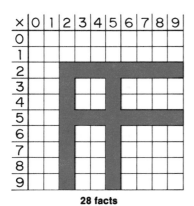

28 facts
Fig. 2.6. Skip Counting

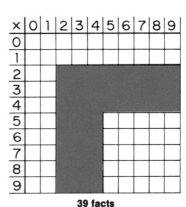

39 facts
Fig. 2.7. Repeated Addition

The *one more set* idea can be used for almost all the facts. If one multiple of a number is known, the next multiple can be determined by adding a single-digit number. This is a powerful strategy. Each fact in turn can be used to help learn the next multiple of either factor. The major difficulties that children encounter with this strategy involve addition where it is necessary to regroup, or bridge a decade. For example, 8 × 6 is 48; so 9 × 6 is 48 + 6, or 54. Many children do not have enough competence to add endings mentally and regroup for this strategy to be effective with those facts.

The gray spaces shown in figure 2.8 indicate the facts that can most easily be solved by this strategy. The others can be solved by adding endings, but each involves regrouping, or bridging a decade. The corresponding addition problems are shown for each of these. The chart also indicates the importance of teaching children to add endings mentally.

×	0	1	2	3	4	5	6	7	8	9
0										
1										
2										
3								14+7	16+8	18+9
4							18+6		24+8	27+9
5								28+7		36+9
6								35+7		45+9
7				18+3			36+6		48+8	54+9
8					28+4			49+7	56+8	63+9
9							48+6	56+7	64+8	72+9

27 facts (49 including those with adding endings)
Fig. 2.8. One More Set

The *twice as much* strategy can be applied to multiples of 4, 6, and 8. An array with one of these numbers as a factor can be split in half. The product is then twice as much as each half.

Some difficulties can occur when the double of a number involves a renaming. For example, it is easier to double 21 than to double 28. Twice as much as 28 is not in the forties, since 8 + 8 is greater than 10. Again, adding endings is a helpful skill.

The gray part of the chart in figure 2.9 indicates the facts that can easily be solved by this strategy. The addition problems indicate other facts that can be solved by thinking twice as much, but where the addition involves renaming.

×	0	1	2	3	4	5	6	7	8	9
0										
1										
2										
3										
4									16+16	18+18
5							15+15			
6						15+15	18+18			27+27
7									28+28	
8					16+16			28+28		36+36
9					18+18		27+27	36+36		

20 facts (33 including those with renaming)
Fig. 2.9. Twice as Much as a Known Fact

Using *facts of 5* can be helpful for any of the problems with large factors. However, it is most helpful for multiples of 6 and 8. In each situation, 5 sixes or 5 eights is a multiple of 10; so it is rather easy to add on the remaining part. For example, 5 sixes is 30, and 2 sixes is 12; so 7 sixes is 30 + 12, or 42. The remaining facts that are not colored in figure 2.10 involve multiples of 9 or 7. Two-digit addition with renaming is required to use this strategy for those facts. There are easier ways to solve them.

Patterns can be used for many of the facts, but they are probably most effective with the multiples of 9. The chart in figure 2.11 indicates this.

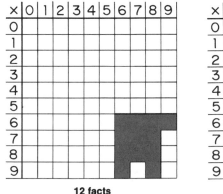

12 facts
Fig. 2.10. Facts of Five

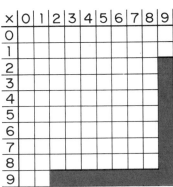

15 facts
Fig. 2.11. Patterns

Together the strategies described above provide at least one efficient way for a student to solve every multiplication fact other than the facts with zero and one. Those can generally be learned from the initial work with multiplication.

Teaching the thinking strategies

Although children are familiar with skip counting by twos and fives and skillful at addition, both skip counting and repeated addition should initially be presented together with a model; they should not replace a model until the children have developed the concepts involved.

Activity: Show three sets of 5.

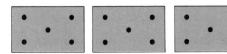

Point to each set as you
count.

"What multiplication problem is shown?" (3 × 5)

"How many dots in all?"

"5, 10, 15.
3 × 5 = 15."

Activity: Write 4 × 2 = _____

"What problem is shown?"

"How could we show this on the number line?"
(four jumps of 2 each)

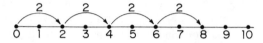

Indicate each jump as you count.

"Let's count by twos. 2, 4, 6, 8. So 2 × 4 = 8"

Activity: Make part of a hundreds chart. Have the pupils circle each multiple of 3. Then let them use the chart to learn to count by threes. Similar patterns can be used to help with multiples of 9.

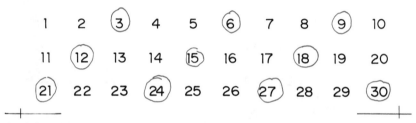

1   2   ③   4   5   ⑥   7   8   ⑨   10

11   ⑫   13   14   ⑮   16   17   ⑱   19   20

㉑   22   23   ㉔   25   26   ㉗   28   29   ㉚

Activity: Write the numbers in four columns. Have the students find multiples of 4. Count by fours.

```
 1    2    3    4
 5    6    7    8
 9   10   11   12
13   14   15   16
```

⋮

Use the chart to count.

"How much is 3 fours?"
"4, 8, 12. So 3 × 4 = 12"

The one-more-set strategy can be developed by means of many different models. Only the set model and the array are shown here. Before this strategy can be effectively used for some facts, the children need to be proficient in adding endings.

Activity: Show 3 sets of 7.

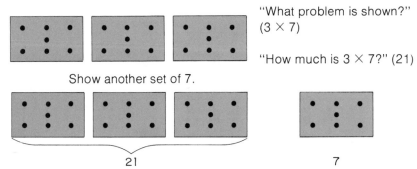

"What problem is shown?"
(3 × 7)

"How much is 3 × 7?" (21)

Show another set of 7.

21   7

"3 × 7 = 21. How much is 4 sevens?"

"21 + 7 = 28; so
4 × 7 = 28."

Activity: Provide arrays that are already split. Ask the children to fill in the blanks.

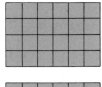

4 sixes is _24_

1 six is _6_
5 × 6 = _30_

Activity: Show 6 × 4

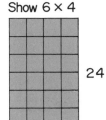

24

"What is 7 × 4?"

"Let's start with a fact we know."

"6 × 4 = _____?" (24)

Show another set of 4.

4

"7 fours is just one more set of four."

"24 + 4 = 28"

The twice-as-much strategy can be developed by splitting arrays in half. It can be used any time there is an even factor.

Activity: Provide arrays that are already split. Ask the children to fill in the blanks.

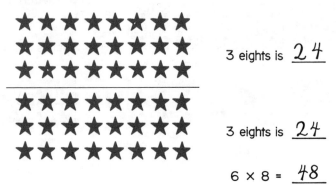

3 eights is $\underline{24}$

3 eights is $\underline{24}$

6 × 8 = $\underline{48}$

Activity: Provide arrays showing multiples of 4, 6, or 8. Have the children split them in half and complete the blanks.

$\underline{2}$ sevens is $\underline{14}$

$\underline{2}$ sevens is $\underline{14}$

4 × 7 = $\underline{28}$

Using facts of 5 can also be developed from an array. Since most children learn the facts of 5 before working on multiples of larger factors, this strategy enables them to use what they already know.

Activity: Provide an array with factors greater than 5. Split the array so that the first five rows are in one part.

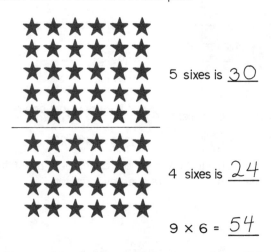

5 sixes is $\underline{30}$

4 sixes is $\underline{24}$

9 × 6 = $\underline{54}$

Besides the emphasis on a specific thinking strategy, there are many general activities that help students develop mature thinking strategies.

Activity: Focus attention on the strategies and not on the answer by having the same array repeated on a page about eight times. Have the students split each array in a different way. After they have completed the task, discuss the different possibilities. Ask which way is the easiest. This activity is particularly effective for 7 × 8.

Activity: Have each student get a partner. Give them a fact problem, for example, 6 × 8. The first student does one part of the problem, for example, 2 eights is 16. The second student must complete the problem—in this case, 4 eights is 32. Then the first child can add the two parts to get the product. Encourage the students to split the product into parts that are easy to add.

An instructional plan

The following steps are one possibility for planning instruction for the multiplication facts. Children are encouraged to develop mature thinking so that they can successfully give quick responses to fact problems.

Step 1. Introduce multiplication using the models described above. Introduce one model at a time.

Step 2. Use a model to develop skip counting and repeated addition to solve appropriate problems.

Step 3. Use models to develop the one-more-set strategy.

Step 4. Use drill and games for facts having at least one factor less than 6.

Step 5. Use the array model to develop the strategies that involve splitting the products into known parts. Develop one strategy at a time.

Step 6. Practice the facts that can be solved efficiently by splitting the product into parts.

Step 7. Use patterns to help children learn the remaining facts.

Step 8. Drill on all the facts to develop speed and accuracy.

This sequence would normally be used over a two-year period in the middle grades. Each step can be roughly interpreted as a unit of instruction.

Older children who have not yet learned all the multiplication facts can also profit from experiences with thinking strategies. Quite often pupils who are having difficulty with some of the harder facts are using only repeated addition. These children may need to be introduced to the array as a model for multiplication before trying to develop more mature "splitting" strategies.

Summary

Children profit from explicit instruction to help them learn thinking strategies for solving basic facts. These strategies should be an integral part of instructional programs for these facts. The framework presented in this article provides a way to organize instruction so that learning the basic facts becomes a meaningful experience.

REFERENCES

Brownell, William A., and Charlotte B. Chazal. "The Effects of Premature Drill in Third-Grade Arithmetic." *Journal of Educational Research* 29 (September 1935): 17–28.

Jerman, Max. "Some Strategies for Solving Simple Multiplication Combinations." *Journal for Research in Mathematics Education* 1 (March 1970): 95–128.

Swenson, Esther J. "Organization and Generalization as Factors in Learning, Transfer and Retroactive Inhibition." In *Learning Theory in School Situations*. University of Minnesota Studies in Education, no. 2. Minneapolis: University of Minnesota Press, 1949.

Thiele, C. L. *The Contribution of Generalization to the Learning of Addition Facts.* New York: Bureau of Publications, Teacher's College, Columbia University, 1938.

3

Games: Practice Activities for the Basic Facts

Robert B. Ashlock
Carolynn A. Washbon

CHILDREN enjoy games! In school and out, they play games frequently, often making up their own rules. Teachers concerned that children learn the basic facts of arithmetic should be encouraged that children *do* delight in playing games, for games have an important role in any instruction designed to help children master and maintain the basic facts. Games should not be just a pleasant activity to be enjoyed after daily written exercises are completed; they should not be just frosting on the cake. Instead, games should be a regular part of instruction in arithmetic.

In this article we shall consider (1) the role of practice activities, (2) guidelines for selecting and preparing practice activities, (3) games as practice activities, (4) characteristics of games that are effective practice activities, (5) examples of such games, and (6) suggestions to help teachers create their own games.

The Role of Practice Activities

Many instructional activities help children learn the meanings of operations on whole numbers and procedures for solving examples such as $4 \times \underline{\quad} = 36$. Children might use manipulatives or tally marks to figure out the missing factor, and instruction might emphasize the relationship between multiplication and division. Such developmental activities, however, are not activities that lead to mastery. Instead, activities for mastering

39

the basic facts of arithmetic—practice activities—are those that focus primarily on the recall of information. In the language of the computer age, they involve the retrieval of data already encoded or stored in the memory.

Obviously, if information has not been stored in a child's memory, then it cannot be retrieved. Children involved prematurely in activities for mastery often resort to guessing. At best, immature procedures are reinforced and fixed. Since it is important that practice activities not be provided prematurely, teachers must know when practice *is* appropriate. Practice activities for the mastery of the basic facts should be provided after pupils have the information stored *in an orderly fashion* in their memory. This occurs when a child has gone beyond a simple association of stimulus and response and developed a conceptual system for relating the information to other information that has been learned.

Some characteristics of children who have the basic facts stored in memory and are ready for practice activities follow:

---

1. When they are given a basic fact of arithmetic, they can state or write related facts. That is, they can generate other facts with the same addends and sum or the same factors and product.

Given	*Sample Responses*
$4 + 3 = 7$. Facts with the same sum are . . .	$5 + 2 = 7$; $6 + 1 = 7$
$8 + 7 = 15$. Facts with the same sum and addends are . . .	$7 + 8 = 15$; $15 - 7 = 8$

2. When an addend (or factor) and an operation are specified, children can make a fact table. For example:

    $$3 + \underline{1} = \underline{4}$$
    $$3 + \underline{2} = \underline{}$$
    $$3 + \underline{} = \underline{}$$
    $$3 + \underline{} = \underline{}$$
    $$\vdots$$

3. When they use a rule, such as one of the properties for addition, they can explain how they got their answer.

4. When they are given an example, they can solve it in two or more ways. For the example $5 + 9 = \underline{\quad}$, children might develop solutions in the following ways:

5 and 5 is 10 . . .		5 and 10 is 15 . . .
Add 4 more . . .	*or*	So 5 and 9 is one less.
10 and 4 is 14.		5 and 9 is 14.

When a child can do such tasks, albeit slowly and deliberately, we can infer that the child has developed an adequate conceptual scheme to justify the use of practice activities for mastery.

A mastery of the basic facts need not be acquired before a child is introduced to computational procedures incorporating those facts. In order to maintain and increase skill, teachers should provide practice activities on the basic facts during and after the learning of computational procedures.

Guidelines for Selecting and Preparing Activities

When a child is ready for practice, the teacher needs to select or prepare activities with care. Practice activities may include the use of flash cards and worksheets as well as games. Some guidelines for the selection and preparation of practice activities follow:[1]

1. *Activities should involve a time constraint.*
2. *Immediate feedback should be provided for the child.*
3. *Activities should have a minimal penalty for error.*

Children participating in practice activities having these characteristics, which are discussed later, are much more likely to acquire the mastery desired.

Games as Practice Activities

Games can be valuable as practice activities when selected carefully and employed judiciously. What makes a game especially useful as a practice activity?

Games offer a highly motivating format for practicing the recall of the basic facts. Their competitive aspect is very attractive to children—provided, of course, they have a reasonable hope of winning. Children do not need to win every time. A game can provide an instructional context in which it is "safe" to be wrong, thereby increasing the likelihood that children will "pull the answer out of their heads," that is, practice recall rather than figure out the answers again. Games also offer children a welcome change of pace from quiet seatwork. Instead of working alone with paper and pencil, they find themselves involved with colorful objects and participating in excited conversation. There is action. Sometimes they win, and winning is exciting. Games offer a unique motivating format for practice activities.

1. The authors are especially grateful to John W. Wilson and the students in the Arithmetic Center at the University of Maryland for the help they have provided.

What makes an activity a game? Games to be used for practice activities are, first of all, games. Teachers sometimes call an activity a "game" when they want to convince children the activity will be fun, but not every activity qualifies as a game. Games consist of an imagined context, materials, rules, a decision-making mechanism, the actions of the players, and a payoff.

Here we shall consider only games used for practicing the basic facts of arithmetic. Imagined contexts can vary widely: athletic contests; stock-car races, space satellites. Materials for games can include boards and other devices on which to play, cards, dice, spinners, and so forth. Games can also be designed to include a math balance, rods, or other instructional aids. The rules of a game govern the actions of the players and their use of the materials. An important feature of a game is a decision-making mechanism, such as a spinner or a bid. This feature gives a game an important source of motivation—the element of chance. And of course the payoff involves reaching some kind of predetermined goal. These observations apply to all games. What else should be considered if games are to serve as practice activities?

Games selected, created, or adapted for use as practice activities for the basic facts of arithmetic should follow the guidelines for selecting and preparing practice activities noted in the previous section. A time constraint should be placed on the moves of the players; if no such constraint is provided, a child is likely to practice figuring out the answer rather than practice recalling basic facts. Egg timers are one means of providing a time constraint. The game should also provide some means of immediate feedback to a player in regard to his or her response; tables and charts often meet this need. Further, any penalty for error should be minimal. Severe penalties, such as having to go back to the beginning or losing all the points earned, reduce a child's hope of winning and hence reduce interest in the game. Games used for practice activities for the basic facts should follow two additional guidelines: (1) the basic facts selected for the game should appear in random order, and (2) the game should provide each player with an equal chance of winning (this is in part a function of the teacher, who should encourage children of comparable skill to play together).

Examples of Games

The four games presented on the following pages are examples of well-designed activities providing practice on the basic facts of arithmetic. They illustrate a variety of imagined contexts, formats, and decision-making mechanisms. The common feature of these games is that the players are required to recall basic arithmetic facts in order to play.

SMILE

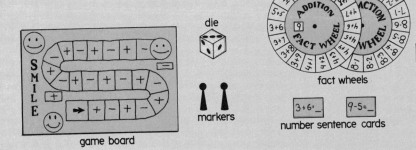

die

markers

fact wheels

3+6=_   9-5=_

number sentence cards

timer

game board

Created by Mary A. Karapetian

Materials: Game board, die, markers, fact wheels (fact wheels can be constructed of lightweight cardboard circles; they are also available commercially), egg timer or stopwatch, and two sets of cards (one set of open addition number sentences, one set of open subtraction number sentences).

Rules:

1. To begin, each player rolls the die. The player with the highest number goes first.
2. The first player rolls the die, moves his or her marker as many spaces as shown on the die, and then draws a card from the addition set or the subtraction set as indicated on the space the marker occupies.
3. Once the card is drawn, another player sets the timer or starts the stopwatch. The player has five seconds to give the correct solution for the sentence on the card.
4. The solution is checked by moving the fact wheel to the appropriate fact and reading the solution in the window.
5. If the player has given the correct solution within five seconds, he or she moves the marker two additional spaces. If the player has given an incorrect solution or does not give a solution within five seconds, no move is made.
6. Play passes to the left.
7. The first player to reach the large happy face at the end of the path is the winner.

Variations: Use multiplication and division number sentence cards and fact wheels. Omit operation signs from the board and color code the cards and spaces.

The imagined context for Smile is the familiar "happy face." The format used is a simple path similar to those found on many game boards. This game is easy to make, and children seem to enjoy playing it. Smile also follows each of the guidelines suggested for games as practice activities:

1. *Facts should appear in random order.* The sets of number sentence cards are shuffled before play begins, and the roll of the die determines whether an addition or a subtraction fact will be drawn.

2. *The game should provide equal chances for winning.* The teacher should encourage children of equal ability to play together to assure each an equal chance of winning.

3. *The game should involve a time constraint.* An egg timer or a stopwatch is used to encourage children to recall facts they have already learned rather than to figure them out each time. The length of time permitted for giving solutions may be varied with the ability of the players.

4. *Immediate feedback should be provided for the child.* Checking the facts on the fact wheel provides immediate feedback for the player giving the solution.

5. *Such activities should have minimal penalty for error.* A player who does not give the correct solution neither loses a turn nor moves back along the path; the child simply does not gain the two extra spaces earned by giving a correct solution within the time limit.

Though Smile was designed as an instructional game for the practice of the basic addition and subtraction facts, it could easily be adapted for the practice of multiplication and division facts by substituting appropriate decks of cards and operation signs on the board.

Beat the Bell, MADS, and Giant Step are other games designed as practice activities. They illustrate a variety of formats, timers, and feedback devices that can be used in games designed for the recall of facts.

An interesting timing device is the pathway used in Beat the Bell. A marble rolls down the pathway in about six seconds. When the detour is removed, the pathway is shortened; then the marble completes its trip in about three seconds. Beat the Bell number sentence cards illustrate a means of providing immediate feedback. The cards are constructed of two pieces of lightweight cardboard taped together with an arrow inserted through the back piece. The correct answer for the example on the front of the card is visible only when the arrow is pulled up.

MADS and Giant Step are examples of active games that can be used as practice activities. In MADS, the distance a player must throw the beanbags can be varied according to the physical skill of the players. Additionally, MADS introduces the use of the low-cost electronic calculator as a means of providing immediate feedback. Either the player throwing the beanbag or another child may operate the calculator.

Giant Step is a team game that can be played indoors or outdoors with a group of children. The large dice used in this game are cut from scraps of foam rubber. Numerals can be put on the dice with a felt-tipped pen or can be cut from very thin pieces of foam rubber and glued on.

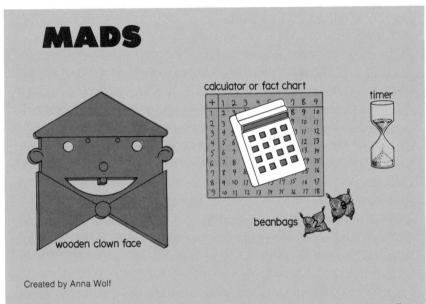

Created by Anna Wolf

Materials: Wooden clown face, calculator or an addition fact chart, five-second timer, pencil and paper, and box of beanbags (two with each numeral 1–9).

Rules:
1. The first player draws two beanbags from the box and tries to throw them through the mouth of MADS, the clown.
2. If the bags go through, the timer is set, and the player must say an appropriate addition sentence using the numbers shown on the bags as addends. If a bag does not go through the clown's mouth, its value is zero.
3. Sums are checked on the calculator or on an addition fact chart by another player. If the sum is correct and given within five seconds, the player scores one point.
4. When all the beanbags have been used, the points are totaled. The player with the most points wins.

Variations: For multiplication facts, players would state an appropriate multiplication sentence using the numbers shown on the bags as factors.

BEAT THE BELL

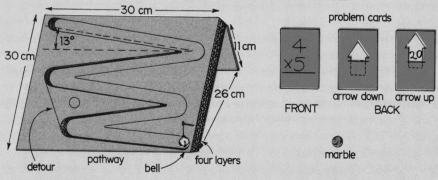

FRONT   arrow down   arrow up

BACK

marble

detour   pathway   bell   four layers

Created by Barbara R. Sadowski

Materials: Pathway with a bell, a marble, forty-five multiplication number sentence cards (facts on the front, arrows with answers on the back), and pencil and paper for scoring. (Construction note: The pathway is constructed of four layers of corrugated cardboard. The path is cut in the top three layers as shown. The fourth layer is glued to the others and a 30-cm-by-11-cm piece of cardboard is taped on as a stand. The detour is made of three pieces of cardboard cut from the path and has a 3-mm wooden dowel or a pin in the center to hold it in place.)

Rules for Two Players:

1. To begin, both players draw a card and pull up the arrow on the back. The player with the higher number goes first. Cards are randomly stacked into three piles of fifteen cards each.

2. The first player takes the top card from any pile and, without turning it over, hands it to the second player.

3. The first player places the marble at the top of the pathway. The second player then holds up the card so that the first player can see the number fact as the marble starts down the path.

4. The first player must give the correct answer for the fact before the marble reaches the bell.

5. If the player beats the bell with the correct answer, one point is scored.

6. Players check answers by pulling up the arrow on the back of the card to show the correct answer.

7. Now the second player chooses a card and play continues.

8. If a player thinks he or she can answer the problems more quickly, the detour may be removed. If his or her answer beats the bell, two points are scored.

9. The game ends when all cards are used. The winner is the player having the most points at the end of the game.

Rules for One Player:

If one player plays the game alone, he or she starts the marble down the path, draws a card, and tries to give the answer before the marble reaches the bell. If the player is successful, the card is kept. If the player is not successful, the card is placed in the board's pile. If the player has more cards than the board when all the cards have been played, the player wins.

GIANT STEP

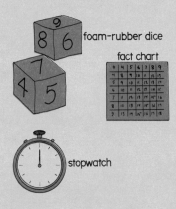

foam-rubber dice

fact chart

stopwatch

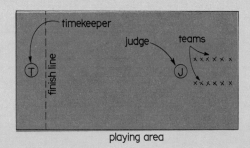

timekeeper

finish line

judge   teams

T

J

x x x x x x

x x x x x x

playing area

Materials: Two large foam-rubber dice with the numerals 4–9 on the faces, an addition fact chart, and a stopwatch or clock with a second hand.

Rules:

1. Form two teams of about ten players each and have them line up at one end of the room or playing area as shown in the diagram. Mark a finish line at the other end of the room. Select a timekeeper and a judge. The timekeeper stands at the finish line, and the judge stands between and just in front of the two teams. Set a time limit of about ten or fifteen seconds.

2. To begin, the judge rolls the dice on the floor and the timekeeper starts the stopwatch. The first player in each line must say the sum of the numbers shown on the top of the dice and take one giant step forward. Each team must move up behind its first player. All these actions must be completed within the time limit.

3. When the time is up, the timekeeper says "Stop." The judge then checks the answers given by the players. If a player has given the correct answer and the team has completed moving up behind the player, the team may move forward another giant step. If the answer is incorrect or if the team has not completed moving up behind the first player, the team does not move the extra giant step.

4. The first player in each line then goes to the back of his or her team's line, and play proceeds.

5. The winning team is the one that reaches the finish line first.

Variations:

1. Use dice with different sets of numerals.
2. Have players say the product of the numbers that appear on the dice.

The interaction of rules, decision-making devices, and materials that characterize Smile, Beat the Bell, MADS, and Giant Step as practice activities for the basic facts is summarized in table 3.1.

TABLE 3.1
Analysis of Games in Relation to Guidelines

Guidelines	Smile	Beat the Bell	MADS	Giant Step
1. Random order of facts	Cards shuffled; roll of die	Cards shuffled	Drawing of beanbags	Roll of dice
2. Equal chance of winning	Random decision-making devices	Random decision-making devices	Random decision-making devices	Random decision-making devices
3. Time constraint	Stopwatch or egg timer	Pathway with bell	Egg timer	Stopwatch
4. Immediate feedback	Fact wheels	Arrows on back of cards	Calculator	Fact chart
5. Minimal penalty	No extra moves if incorrect solution is given	No point scored for incorrect solution	No point scored for incorrect solution or inaccurate throw	No extra steps if solution is incorrect

Creating Games

Games available commercially as well as those described in widely read publications can be used by teachers to provide practice for their students. An analysis of an existing game may indicate that it should be modified if it is to serve as a practice activity. Games for practice can often be improved by adding a time constraint, reducing the penalty for error, or providing a means for immediate feedback. These modifications can usually be made by changing a rule of the game and by using timers and fact charts.

Teachers often wish not only to modify existing games but also to create games that are appropriate to a particular situation. The process of creating a game, especially if it is to serve as an instructional activity, consists of at least two distinct phases—the instructional considerations and the design of the game.

The steps that follow seem most helpful in creating an instructional game. First, make these decisions concerning instruction:

- Select the topic for instruction—for example, the basic facts of arithmetic.
- Identify the purpose of the activity—for instance, practice.
- Determine the behaviors the players should display—recall, for instance.

Then design the game:

1. Select a *format* appropriate for the topic, behavior, and purpose of the activity—a board game, perhaps.

2. Select an *imagined context* for the format—for example, a journey in the jungle.

3. Decide on the *rules* of the game.

4. Select *materials* for the construction of the game.

The following are helpful hints for the construction of games that will heighten interest, increase durability, and reduce costs of construction.

- Use bright colors, pictures, and designs to make attractive game boards. Themes or imagined contexts such as flowers, cartoon characters, or stock cars add interest.
- Use dice, spinners, wheels, or similar devices for determining moves and checking answers.
- Cover all the pieces of a game with the same pattern of contact paper. Store the pieces in a box that is covered with the same pattern of contact paper.
- Make cards 6 cm by 9 cm for ease in handling. Put a design on the back so that all cards in the deck will be easy to identify. Cover both sides of the completed cards with clear contact paper.
- Boards made of tagboard about 46 cm by 56 cm that fold in half are more durable when covered with contact paper. One board could be used with several sets of materials or cards for different activities.
- Blank dice and blank playing cards are available from several school-supply firms.
- Spinners can be constructed from plastic lids, a piece of cardboard, and a brass fastener. Foam-rubber dice can be cut with an electric knife. (And remember, they are quiet!)

A mastery of the basic facts is fundamental to computational proficiency. Games that are carefully chosen, modified, or created offer a unique means of encouraging children to achieve a mastery of the basic facts in an entertaining and instructionally sound setting.

BIBLIOGRAPHY

Arnold, Arnold. *The World Book of Children's Games.* Greenwich, Conn.: Fawcett Publications, 1972.

Ashlock, Robert B., and James H. Humphrey. *Teaching Elementary School Mathematics through Motor Learning.* Springfield, Ill.: Charles C Thomas, Publisher, 1976.

Henderson, George L. *Math Games for Greater Achievement.* Skokie, Ill.: National Textbook Co., 1972.

Johnson, Donovan A. *Games for Learning Mathematics.* Rev. ed. Portland, Maine: J. Weston Walch, Publisher, 1973.

Judd, Wallace. *Games Calculators Play.* New York: Warner Books, 1975.

Kennedy, Leonard M., and Ruth L. Michon. *Games for Individualizing Mathematics Learning.* Columbus, Ohio: Charles E. Merrill Publishing Co., 1973.

Kohl, Herbert R. *Math, Writing, & Games in the Open Classroom.* The New York Review Book Series. New York: Random House, 1974.

National Council of Teachers of Mathematics. *Games and Puzzles for Elementary and Middle School Mathematics: Readings from the "Arithmetic Teacher."* Edited by Seaton E. Smith, Jr., and Carl A. Backman. Reston, Va.: The Council, 1975.

4

Suggestions for Teaching the Basic Facts of Arithmetic

Edward J. Davis

DO YOUR pupils have difficulty memorizing the basic facts of arithmetic? Have you wondered why? Are you continually looking for ways to help them? Do they know their facts one day and forget them the next? Do you feel you have to spend too much time on drill? If some of your answers are yes, you may be interested in some principles (guidelines) that other teachers have found helpful in teaching their pupils to memorize the basic facts of arithmetic.

The ten principles discussed here have been followed by teachers in their classrooms. Their reports indicate that these principles can lead to a classroom populated with pupils memorizing addition, subtraction, multiplication, or division facts with success and enthusiasm. Most of the principles will sound familiar—teachers have been exposed to many of them in psychology and methods courses. Perhaps the crush of teaching responsibilities causes many of them to be overlooked; this is understandable. But they are not impossible to follow. Some of these principles have their roots in psychological research. Summaries and interpretations of this research have been given by psychologists such as Hilgard (1956) and Watson (1961) and by mathematics educators such as Sueltz (1953). Other principles, as well as guidelines for teaching algorithms, are

51

more extensively discussed by McKillip (1974). And some reflect my own beliefs, derived from teaching children and working with classroom teachers.

The Principles

1. Children should attempt to memorize material they reasonably understand

Children should be able to provide both physical and mathematical evidence that they understand an arithmetic fact before they are asked to memorize the corresponding symbolic statement. It would be well to require this evidence in more than one setting. Arranging or counting discrete or connected physical objects, making arrays of dots, using a number line, and making tally marks are examples of different settings (embodiments) that can be used. For example, children give evidence of understanding the statement $5 + 2 = 7$ if they can give responses similar to these:

- Five plus two equals seven is true because if I stand on five on the number line and take two steps forward, I can get to seven.
- Here are five fingers and here are two. Put them together and you have seven fingers.

A child giving the responses above is using two models for addition—"counting on" and "putting together"—as well as demonstrating an understanding of each part of the fact: 5, 2, +, and =. In other words, the child is giving some evidence of understanding the concepts of five, two, add, and equals and a relationship that exists between these concepts.

Understanding is not an "all or nothing" phenomenon. Understanding a number fact involves realizations somewhat different from the realizations related to understanding a concept. Consider, for example, the tasks presented in table 4.1. Note that one does not argue the truth of a concept (item 5)—concepts are not true or false.

Note that table 4.1 does not demand immediate recall. Children should be able to perform *most* of the tasks in table 4.1 before drill is begun on a systematic basis.

2. Have children begin to memorize basic arithmetic facts soon after they demonstrate an understanding of symbolic statements

Manipulative materials (including fingers) are valuable and probably necessary in developing an understanding of facts, ideas, rules, and procedures. Although mathematics instruction frequently begins with manipulative experiences, there comes a time when one is expected to act through the medium of symbols that represent mathematical ideas and

TABLE 4.1
Understanding Number Facts

	Children may be said to understand a number fact to the extent that they can—	Sample questions and concerns to assess understanding of the fact $5 + 4 = 9$
Understand *what* the fact says	1. *Create or recognize embodiments of the fact*	1. Can you use the number line, the rods, the counters, and your fingers to show $5 + 4 = 9$?
	2. *Understand the concepts in the fact*	2. Do children understand the concepts and symbols denoted by 5, 4, 9, +, and =?
	3. *Use the fact in simple exercises*	3. $\begin{array}{r} 5 \\ 4 \\ +\ 1 \\ \hline ? \end{array}$
	4. *Use the fact in simple problem contexts*	4. Make up a story problem that asks you to use $5 + 4 = 9$.
Understand *why* the fact is true and realize its significance	5. *Show the truth of the fact using objects, models, or other facts*	5. Suppose I say that $5 + 4 = 10$. Can you show me I am wrong? Starting with $4 + 4 = 8$, can you show me that $5 + 4 = 9$?
	6. *Complete related statements of the fact*	6. Does $5 + 4 = 4 + 5$? Why? What makes $5 + \square = 9$ true?

procedures. This is a cyclic and sequential phenomenon. A child who has successfully completed an instructional cycle for addition and can meaningfully use symbols to represent addition may still be using manipulative objects in learning about subtraction. Once a child understands the symbols 3, 2, +, =, and ?, he or she is ready to tackle the meaning of combinations of these symbols, such as $3 + 2$, $3 + 2 = 5$, and $3 + 2 = ?$ Physical objects can, and probably should, be used in developing meaning for mathematical symbols.

The point of principle 2 is that *soon* after pupils have acquired an understanding of statements of basic facts such as $3 + 2 = 5$ and $3 + 2 = ?$, they are ready to begin to memorize these facts. The understanding of a basic fact was discussed in principle 1.

If drill before a reasonable understanding has occurred is "too early," is it not also possible to start "too late" in memorizing basic facts? For many children a systematic drill program, beginning with addition facts, should begin in grade 1. All the addition facts need not be memorized before drill on subtraction facts can begin. After children understand the meaning of statements like $4 - 3 = 1$ and $4 - 3 = ?$, they can begin to commit basic subtraction facts to memory. This allows the memorization of the basic facts of arithmetic to proceed at a reasonable pace. It may also help to prevent them from becoming so dependent on their fingers or tally marks that they resist giving up these counting habits, which at one stage were needed in developing their understanding of arithmetic operations.

3. Children should participate in drill with the intent to memorize

Some children scramble through a drill session by figuring out answers. It is not uncommon during drill to see children rapidly counting on fingers or quickly making tally marks and counting. These children are *not* practicing remembering facts. They are practicing finger counting or tallying, and they become skilled at these procedures! Teachers must communicate the goal of committing facts to memory and get children to accept and work toward this goal. Children who figure out answers during a drill session are doing little to improve their memory of basic facts. Principles 4 and 5 can help teachers get children to accept and strive for the memorization of basic facts.

4. During drill sessions, emphasize remembering—don't explain!

Use your precious few minutes of drill time to get children to practice remembering. Don't take time to work out or demonstrate how the correct answer is obtained. Children should have this understanding before beginning a drill program. If you feel a review is called for, either stop the drill session entirely or lead a review later. Don't keep interrupting a drill session to review.

It may seem strange to advise against working out a forgotten fact. But suppose—just suppose—you were memorizing multiplication facts for 17. You are asked to respond to the question "Seventeen times eight equals ?" and you draw a blank—you have forgotten the answer. Do you want, or need, to perform this computation:

$$
\begin{array}{r}
17 \\
17 \\
17 \\
17 \\
17 \\
17 \\
17 \\
+17 \\
\hline
\end{array}
$$

That would be tedious, perhaps demeaning, and give you practice in addition—not exactly what you need at this moment! You know what 17 × 8 means; you want the correct answer—136—and you want it right away! Repeating this fact a time or two would be much more helpful in *memorizing* 17 × 8 = 136 than performing the repeated addition above or using the distributive property to calculate the answer.

When you drill, *drill*—explanations can come at another time. Present the question and the correct response as close together as possible. Verifying the correct response is sometimes called *feedback*. Feedback should occur as soon as possible after each response.

If you hold up a flash card marked

$$4 + 5 = ?$$

and the child doesn't know the answer, turn the card over to show the answer or say it. Get the child to repeat the correct answer once or twice and come back to that child with the same fact in about twenty or thirty seconds. Be friendly, smile, don't get upset at lapses of memory—they happen to everyone.

5. Keep drill sessions short, and have some drill almost every day

A five-minute drill session is just about right; ten minutes should be the maximum. Twenty- to thirty-minute drills have helped drill acquire an unpleasant reputation. Two or three five-minute drills a day will enable children to memorize basic arithmetic facts. One drill can take place at the beginning of the mathematics lesson, another at the close of the lesson. Make use of pauses in the schedule, such as when standing in line to go home, to lunch, or to an assembly, to sneak in a few minutes of drill.

6. Try to memorize only a few facts in a given lesson, and constantly review previously memorized facts

In teaching, you will not be adding new facts (new only in the sense that they have not yet been committed to memory) at every drill session. You will have to judge when a group of children has achieved mastery and is ready to attack new facts. Generally speaking, three or four new facts are a challenge that most students will accept. Naturally this figure can be adjusted to meet the needs of a particular child or group. Everyone forgets—*everyone*. Constant review distributed over all the elementary grades is necessary, even after mastery is attained. Of course every previously memorized fact need not be reviewed in every drill session, and once all the basic facts have been memorized, drill sessions do not have to be daily occurrences.

Here, in a six- or seven-minute drill, is a suggestion for controlling the ratio of new facts to old facts. Assume that you are working with some nine-year-old children on memorizing the multiplication facts 9 × 5 and 9 × 6 and that flash cards will be used. Construct a deck of ten flash cards and have six of the cards display any previously memorized facts. Have the remaining four cards display the two new facts as 9 × 5, 5 × 9, 9 × 6, and 6 × 9. (Consider letting the children construct these new cards.) Discuss the new facts (it would be appropriate at this time to verify the answer placed on the cards). Say the new facts together three or four times; then shuffle the deck and begin drilling. Shuffle frequently to keep rearranging the order in which the facts are presented. Let children take turns presenting cards and choosing or creating ground rules for this session. After two or three minutes, replace the six cards displaying old facts with six more cards displaying other previously memorized facts.

The procedure just described controls the ratio of new facts to old facts. It is also designed to insure that children get correct answers most of the time, since the previously memorized facts outnumber the new ones. Success helps keep the children's interest on the lesson. The commutative property of multiplication was illustrated in the construction of the "new facts" cards. Twelve old facts were reviewed while two new facts were the primary object of the drill.

7. Express confidence in your students' ability to memorize—encourage them to try memorizing and see how fast they can be

All children come to school having memorized hundreds of facts. On sight they can call out the names of toys, foods, people, television characters, commercial products, and slogans that television commercials have drilled into their memory! Many reading programs call for children to develop a sight-word vocabulary whose numbers quickly exceed the total number of basic arithmetic facts. In a discussion with your students, point out the extensive amount of nonmathematical facts they have already memorized. Make flash cards for some of these to prove your point. Capitalize on the pleasure and pride they have in possessing this memorized information, as well as its usefulness, by playing games based on memory. Show how fast they can work computational exercises. Have one child hold up a flash card with a display such as 3 + 2 = ____ or some other arithmetic fact they have memorized. Point out that when using the flash card as a stimulus, they can give a memorized fact much quicker than they can punch the correct keys on a calculator to get the answer!

8. Emphasize verbal drill activities and provide feedback immediately

Many children like to respond verbally in drill sessions. If conditions permit, let them "let off some steam" by responding loudly. Drill can be an enjoyable, "togetherness" activity for the entire class. Sessions where everyone responds simultaneously can be a lot of fun, especially if the teacher can correct dissonant responses in a good-natured way. Many children enjoy drilling themselves. Drills where children monitor their own responses and control the pace of the presentation are sometimes called silent drills. A child can use flash cards, a slide projector, or a computer terminal for silent drill. Surprisingly, children can practice remembering facts while watching television. They use the commercial breaks (if they are not too interesting) as short drill sessions. All they need is some device, such as a flash card, to present the question (9 + 4 = ?) and then to reveal the correct answer so that immediate feedback can be provided.

9. Vary drill activities and be enthusiastic

Anything can become boring if it is always done the same way. Your drill activities don't have to be spectacular—your enthusiasm, your confidence in the children's ability to memorize, and your praise are more important than clever drill procedures. Just letting children control and manipulate an apparatus, be it flash cards, a slide projector, or a homemade aid such as a concentration board, provides variety. Your pupils and colleagues can suggest dozens of ways to use flash cards. The children in figure 4.1 are using a commercial version of a set of flash cards for drill on almost all the multiplication facts. The answer appears as

Fig. 4.1

each key is depressed. Children can use it to drill themselves. A teacher can use tape to mask out rows and columns of facts to have children work only on certain combinations.

An interesting and popular flash-card activity is Quickdraw. Divide the class into groups of threes or fours. Try to place children of approximately equal proficiency in each group. One child holds up a card. The other two "draw" (give their response) as if they were characters in a western movie. The first one to call out the correct answer wins that draw (fig. 4.2). Since characters in westerns carried six-shooters, you could stipulate that one needs to win six draws to win the match. The winner (or loser) changes roles with the student holding the flash cards, and action begins again. Your pupils will undoubtedly think of other ground rules to vary this activity.

Fig. 4.2.   Children playing Quickdraw

10. Praise students for good efforts—keep a record of their progress

Be sure to praise the effort and not the child. This allows you to be consistent when expressing concern or displeasure over lackadaisical effort. Every child needs to feel accepted as a person. And there are times when a teacher needs to convey the message, "I understand your feelings and I like you—but I do not approve of what you have done."

Records help teachers provide appropriate instruction. Records can also serve to encourage a child to strive for progress. With regard to memorizing basic facts, records can be kept on charts for each child. For small groups or individual children, "records" such as the "happy-gram" in figure 4.3 can be sent home to encourage parents to praise their children's efforts.

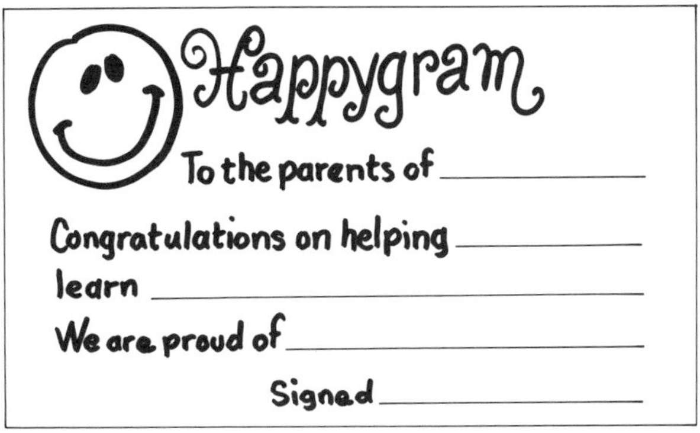
Happygram
To the parents of ____________
Congratulations on helping ____________
learn ____________
We are proud of ____________
Signed ____________

Fig. 4.3

Some Observations

Over one hundred classroom teachers have conducted two- to six-week drill programs following the principles presented here. Teachers gathered pretreatment and posttreatment measures over three- to six-week intervals. They selected time intervals and arithmetic subject matter for drill to fit the needs of particular groups of children or their entire classes. Retention scores were also obtained. The vast majority of these teachers expressed satisfaction with their students' achievement and planned to continue having five- to ten-minute drills almost daily. Although such "action" research does not provide conclusive evidence, it indicates these teachers were satisfied with their results—the morale and enthusiasm of teachers should not be overlooked because of its resistance to precise measurement. It could be that improvements in teachers' morale would frequently result in improved social and academic behavior of students.

A few teachers felt their drill programs were not successful. Discussions with each teacher and written reports and records of each teacher show that in every case where nonsuccess was reported, one or more of the following principles were not followed: 2, 3, 4, or 6.

Principle 2 speaks to the advisability of beginning drill *soon* after understanding is established. Otherwise, children become dependent on finger counting or other manipulative devices and may be reluctant to try to memorize arithmetic facts. They make it difficult for principle 3 (the intent of drill is memorization) to be followed. Such students may require extra encouragement and very careful attention toward engaging them in activities where they are not embarrassed and where they experience success.

When principle 3 was not followed, the cause was always traced to a failure of teachers and students to discuss openly the need to focus on memorization at the outset and throughout the drill sessions.

Some teachers reporting nonsuccess could not bring themselves to follow principle 4 (do not stop to explain). Even though they felt students understood the facts being memorized, some teachers felt they had to stop and explain when errors occurred. These explanations, along with following principle 5 (keep sessions short), undoubtedly took a toll in the time spent on memorizing.

A few teachers apparently advanced at too rapid a pace (see principle 6). Often they spent almost all their drill time on *new* facts; reviews did not characterize their lessons.

This article advances a claim that teaching children to memorize arithmetic combinations does not have to be burdensome, difficult, or distasteful. Many teachers report just the opposite when they follow a systematic program using the principles offered here as guidelines.

REFERENCES

Hilgard, Ernest R. *Theories of Learning.* New York: Appleton-Century-Crofts, 1956.

McKillip, William D. "Teaching for Computational Skill." Athens, Ga.: University of Georgia, Department of Mathematics Education, 1974. Mimeographed.

Sueltz, Ben A. "Drill—Practice—Recurring Experience." In *The Learning of Mathematics: Its Theory and Practice,* Twenty-first Yearbook of the National Council of Teachers of Mathematics, pp. 192–204. Washington, D.C.: The Council, 1953.

Watson, Goodwin. *What Psychology Can We Trust?* New York: Bureau of Publications, Teachers College, Columbia University, 1961.

5

Using Materials and Activities in Teaching Addition and Subtraction Algorithms

Katherine Klippert Merseth

TEACHERS are frequently encouraged to interrelate the concrete and the abstract aspects of school mathematics. For example, the National Advisory Committee on Mathematical Education (NACOME) stated that "current school mathematics curricula can probably be improved by more creative interplay of concrete and abstract ideas" (1975, p. 18). The development of this interplay or bridge from materials to abstractions will not occur automatically. It must be consciously managed and carefully guided by the classroom teacher. This article will outline how this process might be undertaken in the teaching of addition and subtraction in the primary classroom.

As elementary school teachers introduce sets of manipulative materials in the classroom to improve computational skills and make learning them more enjoyable, one common concern arises: how to develop an effective program that will construct a strong bridge from concrete experiences with materials to appropriate algorithms. Regardless of the specific materials used, this concern persists.

Many types of materials are available, some representing the concept of number in a continuous fashion and others emphasizing the discrete

A number of the ideas and methods described for base-ten blocks in this article have been suggested and used by the mathematics coordinator (Arthur J. Short) and teachers of the Newton (Massachusetts) Public Schools.

notion of number. Some materials are commercially produced; others are homemade. There are literally hundreds of materials from which to choose, and the choices are often confusing and difficult.

Base-ten blocks will be used in this article to describe the sequence of moving from concrete manipulations to the abstract algorithms for addition and subtraction. These blocks are only one of many materials that could be selected; several others will be described later. Most base-ten blocks are made of wood or plastic, and some have dimensions in centimeters. The particular set illustrated in this article has four different pieces: units, tens, hundreds, and thousands (see fig. 5.1). The surfaces of all pieces are scored to show their relationship to the unit cube. This helps children check equivalence, since the shapes can always be tested either visually or tactually by counting and matching.

Fig. 5.1. Base-ten blocks and playing board

Whenever any manipulative materials are introduced in the primary grades, ample time should be provided for the children to become familiar with them. As the children experiment with the base-ten blocks, they should be encouraged to discover relationships among the different shapes. This can be accomplished by counting, measuring, building, and comparing. As they become familiar with the relationships, it is helpful to establish a classroom vocabulary to describe the different pieces (children's words have included "units" or "cubes," "sticks" or "longs," "flats" or "squares," and "big cubes" or "blocks"). It is suggested that the introduction of the words *ones, tens, hundreds,* and *thousands* be

deferred until children have themselves discovered the relationships between the pieces and these names. Exploring the trading relationships can
be accomplished individually or in discussions led by the teacher. For
example, "How many flats will it take to make a cube?" Whatever the
answer (an occasional response is six), the child should be encouraged
to build it and compare (see fig. 5.2). If the trading relationships between
the pieces are forgotten, children need depend only on their ability to
count to help them remember the relationships, not on some arbitrary rule
of exchange.

Fig. 5.2. Checking equivalence

Readiness Activities

Once the children are familiar with the materials, they can begin more
formal readiness activities for addition and subtraction.

The Banker's Game

The Banker's Game develops readiness for addition and also includes
grouping by tens (see Davidson, Galton, and Fair [1972]). It is best

played with four or five players and one banker, who tends the bank of blocks. Initially the teacher may be the banker, but children eagerly assume this role after they have had some experience playing the game. Each player begins with an individual playing board like the one shown in figure 5.1. These playing boards are made of heavy, laminated poster-board, approximately 60 cm $\times$ 40 cm, with four columns drawn. It is extremely helpful for the columns to be headed with a drawing of the shape that will be put in that column. Since the columns reflect the place-value system, important readiness skills are being developed for later number work.

Before play begins, the teacher should explain the use of the playing board. Any blocks the children receive must be placed in the column that has the same shape at the top. Thus, all units belong in the column to the extreme right, all sticks belong in the second column from the right, and so on. Play begins in the Banker's Game by a child rolling a die and asking the banker for the number rolled, in *units*. These are then placed in the units column on the child's board. In examining the board, the player considers whether any trades are necessary. The trading rule states that no player may have more than nine objects (units, longs, flats, etc., in the children's vocabulary) in any column at the end of the individual turn. If there are more than nine objects, the child must go to the bank and make an appropriate trade (see fig. 5.3). Teachers may wish to establish a "fun" word to signal that someone's playing board has more than nine objects in a column. Suggestions abound from the children! (We shall use the word *illegal.*) Play does not proceed to the next child until all appropriate trades have been made. The winner is the first player to reach five longs or tens.

As the children gain experience in trading, the game can be extended by using two dice and increasing the winning stakes. To play a continuing game from day to day, the children simply make a record of the blocks they have either by writing the number or by drawing the amount. Later, they can resume the game by representing that number with blocks on their individual playing boards. In fact, children at the third-grade level have continued the game for weeks while playing for the "thousand" cube.

The Banker's Game provides rich opportunities for teachers to develop other number experiences. During the game, the teacher may encourage children to describe their playing boards. The players may also be asked who is closest to winning and urged to compare playing boards in order to develop concepts of order. Children enjoy speculation. Could any player win on just one roll of the die? Two rolls? Children should be encouraged to develop their visual perception. If a fifth column were included on the playing board to the left of the big cube (or thousands) column, how would the blocks look that would belong there? Could the class

Fig. 5.3. Making a trade

build a block to fit the description? Good open-ended questions and discussion stimulate children's curiosity and encourage creative thinking.

The Tax Collector's Game

A readiness game for subtraction with particular emphasis on the regrouping of tens is called the Tax Collector's Game (Davidson, Galton, and Fair 1972). It is simply the reverse of the Banker's Game—the child must now give back in *units* to the bank whatever is rolled on the die. To begin, all players place the same number of blocks on their playing boards. For younger children, this might be three longs and three units; for older children, one flat, one long, and one unit. A child's turn consists of rolling the die and giving back to the bank in units the number indicated. Before passing the die to the next player, however, the child must check to see that all exchanges have been made. The winner is the first player to have an empty playing board. (Teachers should determine beforehand whether an exact roll to go out will be required or not.)

When a child does not have a sufficient number of units (or ones) on hand in the units column to give up, the child will need to "go to the bank" and make a trade. Younger children will occasionally become concerned at this juncture because the playing board is temporarily made "illegal" after the trade until the correct number of units is given up to the bank. For this reason it is advisable to review and emphasize two points: (1) that the rule of having no more than nine objects in any column is important at the *end* of a player's turn and not *during* it, and (2) that no player's turn is over until all appropriate exchanges have been made. Once the rules are reviewed and experience is gained, a few children will quickly begin to suggest alternatives to going to the bank for a trade. For example, when asked to give up six, a child may hand the banker a long (or ten) and ask for four units in return. This reasoning shows some number fact control and signals that the requirement for returning the number in units may be relaxed for that child. Children seem to reach this level of reasoning at widely differing stages. The more methodical students may need specific encouragement to join the others in using the "shortcut." After the children become more comfortable with the game, two dice and larger starting numbers can be introduced to extend the length of play.

Record Keeping

While children are playing the games, it is not suggested that any formal record or score keeping take place. It is important for children to experience the concept of regrouping without being concerned about how to record it. Instead, the teacher may undertake separate activities for developing these skills. For example, the teacher may call out a number, say, twenty-three, and ask the children to represent this number with blocks on their playing boards. After they have represented it, each child could tell how he or she showed twenty-three. Teachers should watch carefully to see how children are representing the numbers and remind them of the "no more than nine" rule. Additional activities with small index cards drawn as miniature playing boards can be developed for visual presentations.

A more advanced series of activities involves the children in recording numerals (with washable pens) directly on their own laminated boards.

For example, the teacher might represent a number on a miniature playing board made from a small index card and ask the children to represent this number with blocks and record it on their own individual playing board.

If a child records two numerals in one column, questions about how many objects are in that particular column may prove helpful. If some children seem to have difficulty both representing with blocks and recording the number on one playing board, it may be helpful to set up two playing boards side by side. One board would be for recording the number and the other for physically representing the number in blocks (fig. 5.4). Using cards with the digits 0 through 9 can also be effective. Depending on the age and reading level of the children, the playing board may be developed further by using the words *thousands, hundreds, tens,* and *ones* to describe the columns rather than the picture of the shapes. With this type of playing board, the reading and writing of more difficult numbers such as 304 (children often write 3004 on hearing this number) can be developed.

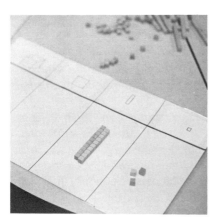

Fig. 5.4. Representing and recording numbers

Some children need a great deal of practice with the recording process. Activities can be developed using cards with numerals written on one side and pictures of the physical representations in blocks on the other side. Games like rummy and Concentration are easily developed by making appropriate playing cards using both block pictures and numbers. Rubber stamps of tens and ones also provide many opportunities for activities. These can be either purchased or homemade. (One teacher cut gum erasers with a pen knife to the appropriate designs and affixed them with glue to blocks of wood to form handles.)

Because the relationships among the different shapes of the blocks are clearly represented, the concept of place value evolves easily and natural-ly. Reversal problems frequently found in the early grades—a child saying "eighteen" and writing "81"—seem to be reduced when this approach of building and recording is employed. If the number is repre-sented correctly with the blocks, the structure of the blocks will not permit reversals in the abstract process of recording. This is because the nu-meral written in a column indicates the number of block pieces present in that column. In this way, the materials provide an often desired but fre-quently missing model of place value in the child's environment.

Addition

As children become more comfortable with representing numbers with the base-ten blocks, their introduction to the addition algorithm can begin. Initially, problems should be presented orally by the teacher and carefully chosen so as not to require regrouping or to contain numbers that seem unrealistically large to the child. For example, the teacher might ask the children to build twelve on their playing boards and then—underneath and separate from the twelve—to build twenty-three. After checking to see that they have represented the two numbers correctly, the teacher would then ask, "How many blocks are there altogether?" The teacher may suggest to the children that they physically gather the blocks to-gether at the bottom of the columns to reinforce the concept of joining together.

After a brief exposure to addition, the children can begin to record the numerals as they physically represent the quantities and then to record their totals. At this time it becomes necessary to introduce and discuss the plus symbol and the line often drawn under the list of numbers. Frequently children think of this line as a "boundary line"; when blocks are joined together, they are brought below this line to show the problem has been completed. (See fig. 5.5).

Fig. 5.5. Stages of solution

Once an addition example has been satisfactorily completed, it is effective for the children to practice reading the problem aloud with all the blocks removed. This gives experience in verbalizing number ideas, using correct vocabulary for mathematical symbols, and visualizing and understanding the example without the blocks. As children become more proficient, examples with more than two addends can be presented, as well as examples in the horizontal form. Teachers might give pupils either just the answer to an example or a partial picture of a completed exercise and ask them to determine what the example might have been (fig. 5.6). Also, so-called word problems could be presented for solution with the blocks. Many of these ideas as well as others can be incorporated in a series of activity cards.

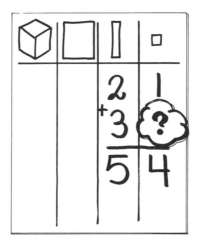

Fig. 5.6. Find the missing number under the cloud

If children have had sufficient experience with the Banker's Game, the move from addition without regrouping to addition with regrouping using the blocks will not be difficult. In fact, many children will ask for "trading problems" or "one where we trade" because they find them interesting and fun. It should be noted, however, that as in the previous discussion on addition without regrouping, examples must be selected with care and forethought. Also, it is very helpful for teachers to present problems within the context of the children's world rather than simply giving them numbers to add. Activities involving the writing of "number stories" or "number headlines" can be exciting and will encourage creativity.

To give an example of an addition problem with regrouping, the teacher might present the following headline: "19 dinosaurs captured in Prehistoria land yesterday! 23 more were captured today! How many dinosaurs were captured altogether?" To do this example, the teacher might ask the children to write 19 and 23 on their playing boards and then to represent these numbers with the blocks.

As the children begin to join the sets, teachers may prefer to encourage them to gather in the units column first. In the given example, there are twelve units altogether in the units column; thus a trade is necessary. (If a child does not notice that a trade is required, a review of the "no more than nine" rule would be helpful.) After the children make the trade, it is suggested that they hold the ten-stick received in trade in their hand at the top of the tens column.

The children next combine all the blocks in the tens column, including the one in their hand, and check to see if a trade is necessary. In this situation no additional trade is required; thus the board is deemed legal, the example is complete, and the result is recorded. The stages of solution are shown in figure 5.7.

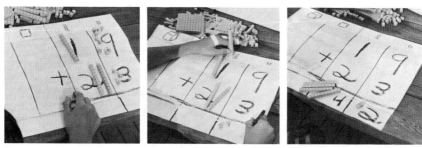

Fig. 5.7. Setting up the example, regrouping, and recording

Although it may seem somewhat unusual to hold the extra ten-stick in the hand, the alternative of placing the piece directly on the playing board after the trade tends to confuse some children, possibly because the "extra ten" is not part of the originally recorded example. It apparently appears to the child as an extra, unaccounted-for object on the board. Also, if teachers prefer to have children record the "carry," or regrouping, number, it may be readily related to the ten held (or "carried") in the child's hand.

It is suggested that the initial experience in regrouping be limited to situations where it is necessary to regroup only in the ones column. Then, as experience is gained, the process may be generalized into the tens and hundreds columns. Also, once the children are comfortable with the trading process, teachers may encourage them to record in each column as they proceed from right to left. In this way the algorithmic process of addition will be more closely imitated.

Subtraction

The subtraction algorithm is more difficult for many children and hence must be presented very carefully. One method for the child's formal introduction to the process of "taking away" with the blocks would be through simple, oral presentations. For example, the teacher might ask the children to represent forty-three on their boards. When this is completed, the teacher would ask them to remove twelve units and then ask what is left or how many are remaining. In this way the child is encouraged to think of subtraction as the process of removing something from the original set. As in the addition examples mentioned previously, the oral subtraction exercises should be presented to children through real-life stories. Encourage them to make up stories for each other. It is often fun and aids in the development of language as well as mathematical concepts.

After children have practiced subtraction on their playing boards without record keeping, the next step is to introduce a recording procedure. For this, the teacher may present the total, or starting number, either visually or orally and ask the children not only to represent it with blocks on their playing boards but also to record the number. Next, the number to be removed is presented, and the children *record* it on their playing boards; it is suggested that they do *not* represent it with blocks. Rather than being represented as a distinct (disjoint) set, the subtrahend is treated as a direction, or a command, to the problem solver. Just as in the oral presentations, this number indicates what part (or subset) of the original set (or minuend) is to be removed (see Sealey et al. [1961]). The child would then complete the example by removing the blocks and recording what remains. Teachers may wish to provide a loop of yarn or string to the side of the original playing board where the removed blocks are placed and also a loop on the playing board around the written number to be removed. In this way instruction and experience in checking subtraction examples can be facilitated.

After the children become more comfortable with the process of "taking away," regrouping, or borrowing, may be introduced in the ones column. More extensive and difficult examples will evolve naturally—particularly if children are encouraged to create their own stories. It is suggested, however, that any child working on regrouping examples in subtraction be experienced at playing the Tax Collector's Game. In fact, a quick review of the game may be very helpful.

In order to examine the regrouping, or borrowing, process more carefully, consider this example:

> The children at North School put 34 gerbils in their cages on Friday after school. When they returned to school on Monday, 19 gerbils had escaped! How many gerbils were left in their cages after the weekend?

After the children examine this situation and represent it with blocks, their playing boards might resemble that of figure 5.8. When they begin to manipulate the blocks, they will look at the units column for nine units to be removed. Only four, however, are currently available. As a result of encountering similar situations while playing the Tax Collector's Game, children will know to go to the bank and make a trade of one long, or ten-stick, for ten units. These ten units are now placed on the playing board, making a total of fourteen units in the units column and two longs (or ten-sticks) in the tens column.

Fig. 5.8. How many were left?

As the trade is made, some children will express a need to record this change on their playing boards. If so, the notation used in figure 5.9 is a possible suggestion. After the trade, the children remove the nine units and next remove the one long or ten. This removal represents escaped gerbils (!), and those blocks remaining are recorded to show the number of gerbils left.

Although it is not necessary, teachers may want to encourage children to work from right to left, always beginning their consideration of the example in the units column. A further refinement would have the children record the results in each column as they proceed. Both suggestions can help imitate more closely the paper-and-pencil process. Care should be taken, however, that the mechanical procedures with the blocks or any manipulative materials do not become overly complicated. If they become too cumbersome and restrictive, many benefits from their use will be obscured.

Fig. 5.9. Recording with subtraction

Further Abstraction

As children gain experience with addition and subtraction examples, the bridge from materials to algorithms must be completed. Much of the groundwork has been laid, but the process of moving from the concrete to the abstract still necessitates careful management and development. The transition is not magical, nor does it happen automatically. It is the result of conscious planning by the teacher. Frequently this planning and management must be keyed to the children's individual needs and learning styles, since the rate at which they move from the concrete to the abstract seems to vary widely. Children who are able to describe what they would do with the blocks without physically moving them would probably be able to proceed through the following sequence of steps toward abstraction with ease and speed. Others who have difficulty visually or verbally reproducing the representations for various numbers may need to progress very slowly and with constant recycling and reinforcement.

A possible first step in removing the block representations and cues is to present addition and subtraction exercises on small cards *without* the shape or name headings above each column. The dimensions of these cards should be consistent with those used previously for visual presentations. An example is given in figure 5.10. If necessary, children may represent the example on their own large playing boards (with headings) and find the answer.

Fig. 5.10. Card without headings

A more abstract stage is to give the children examples on clear acetate sheets. The numerals should be written so that if the sheet were placed directly on a child's large playing board, the numerals would fall in the appropriate columns. Columns may be included on the acetate sheets with dotted or light lines. The child should be asked to do the example directly on the acetate sheet without the benefit of playing boards or blocks. If there is some difficulty, the sheet may be placed directly on a playing board and the example built with blocks and solved. After the solution is determined and recorded on the sheet, the acetate may be removed from the playing board and the situation discussed. For example, can the child describe a mental picture of what the block representations of the numbers look like? Teachers may need to spend more time with some children just moving the sheet back and forth, trying an example with and without blocks, discussing the results and generally supporting the learner in the first attempts to think of, but not physically use, the blocks.

Abstracting somewhat further, examples may also be presented on large-square graph paper, such as in figure 5.11. If a child is unable to begin the example without any cue, the familiar shapes can be drawn in at the top of the columns and the columns darkened as in figure 5.12.

Children should be encouraged to "think of the blocks" and to "picture what the example would look like on your own playing board."

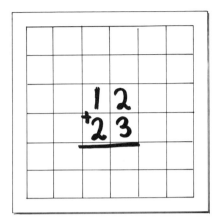

Fig. 5.11. Large-square graph paper

Teachers may even help the child mentally perform the manipulations by acting out the processes with make-believe blocks. Another step toward abstraction would include moving to still smaller graph paper (keeping in mind the limits of a child's fine motor coordination) with no headings or columns darkened. Also, regular lined or ruled paper can be turned on its side to provide a minimal structure for the paper-and-pencil tasks of computation.

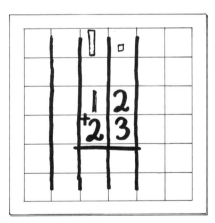

Fig. 5.12. Graph paper with cues

Other Materials

In addition to base-ten blocks, many other manipulative materials can be used to develop the transition from the concrete to the abstract. Commercial materials that teachers may already have available can be adopted to this procedure. For example, Unifix or Multilink cubes can be interlocked to make ten-sticks, and by taping groups of ten-sticks together (with clear tape so that counting is still possible), hundred-squares can be made. Cuisenaire rods can also be used, or Cuisenaire kits that already include the appropriate shapes can be purchased.

Homemade models can be developed from readily available materials. For example, placing ten beans on a tongue depressor and pouring glue over the beans creates a representation for ten (see fig. 5.13). Rafts of these ten-sticks can represent hundreds; loose beans represent ones. Wirtz (1974) provides excellent activities with bean sticks. If tongue depressors are difficult to obtain, Popsicle sticks can be used with small pieces of felt, sandpaper, or adhesive dots to represent the units. In some respects these are more desirable, since the rafts representing hundreds would be smaller.

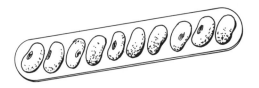

Fig. 5.13. Bean stick

Other materials can be constructed that are slightly different from the model of the base-ten blocks. For example, washers, bottle caps, or similar objects can be collected (the only requirement for these objects is that they all be the same shape and size). In order to represent the tens, use small plastic bags to hold ten objects and close with a wire twist. (It is important that the bags be clear so that children can see how many are inside.) To represent the hundreds, use larger plastic bags to hold the tens-bags. Teachers have also used toothpicks grouped in bundles of tens and hundreds with rubber bands or wire twists. Although toothpicks and rubber bands are attractive because of their availability, younger children may have trouble handling the small toothpicks and the tricky rubber bands. In this situation, Popsicle sticks may be preferred.

The general procedure outlined with the base-ten blocks can be used with only minor variations for all these commercial and homemade materials. Different vocabulary words should be developed to describe the playing pieces, and the playing boards will need the appropriate shapes drawn at the top of the columns.

If a discrete model such as washers, bottle caps, or toothpicks is used, one small change in the procedure can be introduced. In the regrouping process, instead of going to the bank for a trade as with the base-ten blocks, the child may now personally regroup the materials. With addition, for example, the teacher need only have an empty bag ready to give the child when regrouping becomes necessary. If there are more than ten objects in a column, the child personally groups ten objects in a bag and places the bag in the appropriate column. Similarly with regrouping or borrowing in subtraction, the child physically "borrows" a bag, opens it, and uses the contents in the appropriate column.

Conclusion

The examples of materials given here are only a few of the many varieties available. Whether commercial or homemade materials are used, children who have had some or all of these experiences are provided with a physical model of the place-value system. Thus when a child writes 70020 for 720, there is a physical representation that can be referred to and checked. In the same way, reversals—writing 34 when 43 is dictated—can be explained with a concrete model.

Several advantages in the use of base-ten blocks or similar materials have been described. They also provide an additional memory cue. This is particularly helpful for children who have difficulty memorizing numerical processes. Rather than asking the teacher to give an example of "how it goes," the child has other alternatives to help cue his or her memory. The materials can provide a strong and consistent model that can be visualized and considered. In this way the structure of a bridge has been built. Some children will never need to return to the initial, concrete material side of the bridge. Others, however, will constantly need to cross back over and mentally check their ideas and work in more familiar concrete settings. Although this building of a bridge from concrete materials to abstract algorithms is never automatic, it is, with conscientious effort, possible to do and seems to aid many children in the learning of addition and subtraction algorithms.

REFERENCES

Davidson, Patricia S., Grace Galton, and Arlene Fair. *Chip Trading Activities*. Fort Collins, Colo.: Scott Resources, 1972.

National Advisory Committee on Mathematical Education (NACOME). *Overview and Analysis of School Mathematics, Grades K–12*. Washington, D.C.: Conference Board of the Mathematical Sciences, 1975. (Available from the National Council of Teachers of Mathematics.)

Sealey, L. G. W., and others. *The Dienes M.A.B. Multibase Arithmetic Blocks*. Leicester, England: The National Foundation for Educational Research in England and Wales, 1961.

Wirtz, Robert. *Drill and Practice at the Problem Solving Level*. Washington, D.C.: Curriculum Development Associates, 1974.

6

Computation: Implications for Learning Disabled Children

John C. Moyer
Margaret Bannochie Moyer

A S A result of the trend toward mainstreaming—putting children with special needs into regular classrooms—the teaching of computation has become more challenging than ever. Although teachers have always had some of these students in their classes, mainstreaming is increasing the number of such children within the regular classrooms. Primary-level teachers are now teaching addition and subtraction to very heterogeneous groups of children; hence it is important that their professional expertise extend to a practical knowledge of ways to help children with exceptional learning problems. Of particular concern are learning disabled[1] chil-

---

1. The National Advisory Committee on Handicapped Children has provided the following definition of learning disabled children (1968):

Children with special learning disabilities exhibit a disorder in one or more of the basic psychological processes involved in understanding or using spoken or written languages. These may be manifested in disorders of listening, thinking, talking, reading, spelling or arithmetic. They include conditions which have been referred to as perceptual handicaps, brain injury, minimal brain dysfunction, dyslexia, developmental aphasia, etc. They do not include learning problems which are due primarily to visual, hearing, or motor handicaps, to mental retardation, emotional disturbance, or to environmental disadvantage.

We would like to thank Beverly A. Cable for her assistance in photographing the instructional materials.

dren, since they comprise the group most likely to be mainstreamed for a major part of the school day.

Even if learning disabled youngsters are not mainstreamed for all academic areas, the classroom teacher will probably be faced with the task of providing them with some special instruction in addition and subtraction. In her survey Moyer (1975) revealed that mathematics instruction receives low priority among specialists in learning disabilities. (The teachers in the survey ranked mathematics remediation fifth out of seven in terms of instructional time alloted, even though 50 percent of their students had disabilities in the area of mathematics.) Hence, whether learning disabled children are completely absorbed into the regular classroom or whether they are aided by a learning disabilities specialist on a part-time basis, the classroom teacher will probably assume a major responsibility for teaching them to compute.

Some learning disabled youngsters will find computation easy, but others will, for a variety of reasons, fall further and further behind. Some of these children will do very well in oral drill of basic facts but poorly on written assignments involving the same facts. Others will be able to compute problems on paper but will be unable to explain what they have done. Still others will be able to compute an entire page of problems in a workbook but will be unable to do so when the same problems are on the chalkboard. Such irregular performances can be a source of puzzlement for the teacher, especially in light of the fact that learning disabled youngsters are of average or above average intelligence.

Often this puzzlement stems from certain assumptions that teachers unconsciously make about their pupils. Most teachers automatically assume that their pupils possess the underlying processing abilities necessary to learn a computational skill if the skill is viably presented at the children's developmental level. For example, it is usually assumed that Johnny is able to attend to the teacher's instruction or that if Susan can see the problems on the board, she will be able to copy them on her paper. Thus a teacher's attempt to provide adequate instruction usually focuses on whether the child has the prerequisite computational skills rather than on whether the underlying processing abilities are intact. For the majority of students, skill analysis alone leads to a successful instructional strategy, but it cannot be assumed that learning disabled children possess the basic facilities necessary to learn computational skills. In addition to focusing on prerequisite skills, then, the teacher must determine if any of the processing abilities that may affect the pupils' computational performance are lacking.

This article will deal with some of the assumptions that primary teachers usually make regarding the processing abilities their pupils use when learning to add and subtract. Each assumption will be discussed in terms of learning disabled children, and suggestions for the teacher will be offered. It must be noted at the outset that no learning disabled child will exhibit all the disabilities discussed. It is the teacher's responsibility first to determine which of the disabilities a child exhibits and then to apply appropriate instructional techniques.

Attention Span

Assumption 1

Students are able to listen carefully to the teacher's presentation of a concept. Although this is a valid assumption for most children, it is not true of many learning disabled students (and of some normal children as well). No matter how much these children would like to pay attention to the teacher, they have extreme difficulty doing so. Special challenges are in store for teachers who instruct children with this problem.

When teaching computation, an obvious strategy for helping the child with a short attention span is to keep all verbal instruction to a minimum. Brevity alone is not enough, however, for even the briefest of presentations is doomed to failure unless it is made in a clear, logical manner. Thus the careful planning of instruction is of paramount importance. Such planning would include the identification of all sequential prerequisite skills and the elimination of all tangential, extraneous material. Further, the use of colorful visual aids (flannel board, colored chalk, pictures, charts, etc.) peppered with interesting questions and the frequent use of the children's names ("Now, John, as you see, we have two groups of five here") goes a long way toward maintaining attention.

Although these suggestions are sound, there is really no substitute for actual student involvement during the teacher's presentation. For example, when the teacher is first teaching the concept of addition, each pupil could be provided with a set of cardboard disks. The children work along with the teacher, counting out, for example, a set of five and a set of three as the teacher does so on the flannel board. When the sets are joined on the flannel board, the children do the same at their seats, counting the total number along with the teacher. Of course, the teacher must watch individuals to make certain they are all performing the correct actions.

Assumption 2

Students are able to concentrate on assignments when independent work is required. Gaining proficiency in computation can be especially difficult for learning disabled children with short attention spans, since much independent practice is required if the processes involved are to be mastered.

These children have difficulty keeping their minds on a task for any extended period of time because they are easily distracted. Hence, it is imperative that distractions be kept to a minimum. A simple rule would be to require students to keep their desks clear except for the material they are working on. This would alleviate the temptation to play with rubber bands, rulers, and so forth, at a time when their computation exercises should be the main focus of attention. Often these students will work more efficiently if they are allowed to move their desks to face a wall or behind a divider where distractions again will be minimized. In fact, distractible children will frequently choose this option because they realize they will accomplish more work in their "private offices" where their classmates and other class activities will not bother them.

Since the learning disabled child may be able to work efficiently for short periods of time, it is far better to provide many short assignments rather than one long one. If a worksheet is assigned, for example, it could be cut into parts to allow the child to do three separate assignments with four problems each rather than one assignment of twelve problems. The worksheets can be placed in work boxes labeled "To Do" and "Done" so that the child can independently obtain new work as each assignment is handed in. The movement involved allows the student to work off pent-up energy and provides the break such children need to maintain the required attention. Further, the child is given a feeling of accomplishment as each sheet is completed rather than the feeling of defeat that inevitably accompanies an assignment that the youngster considers to be insurmountably long. Gradually as the child's ability to complete assignments increases, the number of problems in each assignment could be increased.

Assignments that entail concomitant motor activity will often guarantee more complete involvement on the pupil's part. Manipulative materials (such as multibase arithmetic blocks, Popsicle sticks, colored rods, and attribute blocks) serve this purpose well.

Attention can also be increased by encouraging the child to talk through the problem on which he or she is working. For the child with a short attention span, being allowed to talk aloud (quietly!) while doing an assignment serves several purposes. First, the self-direction provided by the sound of the child's own voice helps maintain the child's attention on the task at hand. Second, as implied by the verbal mediation studies of Vygotsky (1962) and others, the youngster's thoughts often become more clearly organized if he or she talks the computation process through step by step. Third, the teacher is aided diagnostically because the child reveals the thought processes being used.

Visual-Motor Ability

Assumption 3

Students have sufficient eye-hand coordination to manipulate objects and write symbols. Many of the learning disabled children being mainstreamed into regular classrooms exhibit a disability in visual-motor performance that can interfere significantly with their mathematical functioning. Initially this type of disability may be evidenced in the child's inadequate manipulation of objects. The youngster often displays a noticeable lack of precision as he or she attempts to place counters in groups. The result is that some counters end up on the floor while others lie in such disarray that the child has no way of determining which ones have been counted and which ones have not. Certain modifications are necessary to help such a youngster benefit from the use of manipulative materials.

One suggestion is to string Styrofoam or sponge balls on a rope with knots tied on each end (see fig. 6.1). The child can then push these counters together to form sets of varying cardinality. The advantages of this manipulative for the child with eye-hand coordination problems are clear: (1) the tactile feedback provided by the counters as they are pushed along the rope helps the child coordinate visual and motor processes; (2) the linear nature of the rope provides a ready structure for the organization

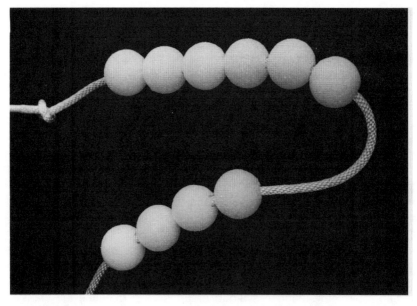

Fig. 6.1

of the counting; (3) the friction between the sponge and rope keeps the balls from moving out of place, even when dropped; (4) the balls make little noise as they hit the floor, and so other children are not distracted.

For those youngsters who have developed greater fine motor control but still have difficulty maintaining groups of counters, a conventional desk-top ink blotter can be a helpful device. Substituting a piece of felt for the blotter produces a miniature "flannel board." The child will find that felt "counters" being counted or arranged in groups will adhere well to the board. Also, the child can move the board without disturbing the arrangement of the counters.

Inadequate eye-hand coordination can also affect children's writing; they often experience problems relating a visual image to the pattern of movements necessary to copy it. Such children have difficulty forming numerals even though they may recognize them and understand their meaning. This difficulty is somewhat analogous to the normal child's inability to draw a chair even though a picture of a chair is easily recognized. Children having this difficulty need visual-motor training. Such training might include having them trace numerals formed with yarn or made out of coarse-textured material (fig. 6.2). This procedure helps them "feel" the configuration of the numeral while looking at it. Models with directional cues help the child determine where to start forming the

Fig. 6.2

numeral (fig. 6.3). Until the child attains some proficiency in writing numerals, ways of circumventing this problem should be devised so that mathematical development can proceed as normally as possible. For example, the child might perform computations with cutout numerals on a flannel board. Rubber stamps of the numerals zero through nine could be provided for recording the answers, or the answers to problems could be formulated in a multiple-choice format so that the child need only circle the correct answers.

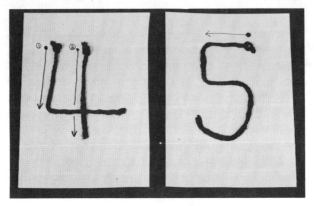

Fig. 6.3

A child with a visual-motor disorder may have high aptitude for mathematics, even though he or she performs poorly on standard computational problems. This disorder (as well as the visual-spatial disorders discussed under assumptions 4 and 5) need not preclude conceptualization in mathematics; however, it may interfere with the *mechanics* of mathematics. Hence, when computational activities are planned for these children, careful attention should be given to methods of reducing the mechanical demands made on the pupils so that they are freer to concentrate on the concepts involved.

Visual-Spatial Ability

Assumption 4

Students have the spatial abilities necessary to copy and organize their assignments into formats that are readable and perceptually workable. Some children have difficulties organizing problems on a page. They may copy the problems randomly and orient them erratically using the nearest available space. The sizes of their numerals may vary greatly for no apparent reason. Numerals may overlap one another, making them difficult to read. Proper alignment of the digits for problems in addition and subtraction may be ignored. (See fig. 6.4.)

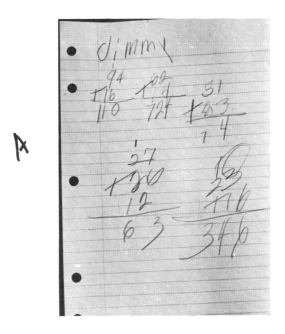

Fig. 6.4

For the learning disabled child having difficulties with visual-spatial organization, this disorganization is not due to carelessness. No amount of pleading, cajoling, threatening, or bribing will ameliorate the problem. Rather, the child must be given aids to help him or her maintain the organization necessary for accurate work.

One suggestion is that the child do as much work as possible in a workbook or on handouts rather than from a hardback book. This greatly reduces the amount of organizational skills required, since the problems do not have to be copied on a separate page. As a further aid, color "highlighters" can be used to help the child line up answers:

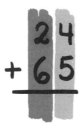

For problems involving regrouping, boxes can be used very effectively:

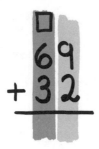

Here the child also receives help in lining up the regrouped units over the tens column. Lastly, when it is necessary for the child to copy problems from a text, the use of large square graph paper or regular lined paper rotated 90° helps the youngster line up the numerals.

The practice of using workbooks and devising worksheets to reduce difficulties stemming from copying may lead to a different problem. Sometimes a child becomes so overwhelmed by the number and arrangement of the problems on a page that he or she knows neither where nor how to begin. To avoid this difficulty, care should be taken to put a minimal number of problems on a given worksheet and to leave adequate space between them. When workbooks are used, the child can be helped by paperclipping window cards to the pages to reveal only a limited number of problems (see fig. 6.5). After the child has completed these, another window card can be positioned to reveal other problems.

Assumption 5

Students can consistently distinguish between the right and left sides of a multidigit numeral. Difficulties in right-left orientation can cause specific problems when the concept of place value is introduced. In spite of an assumption to the contrary, many children simply cannot see any difference between numerals like 21 and 12. Since verbal instructions, such as "The tens are recorded to the left of the ones," are meaningless, visual cues may be necessary to help the children distinguish between multidigit numerals and their reversals. Color coding is often instructive: charts can be made indicating that a blue numeral represents units and an orange numeral represents tens. Thus 43 would be repre-

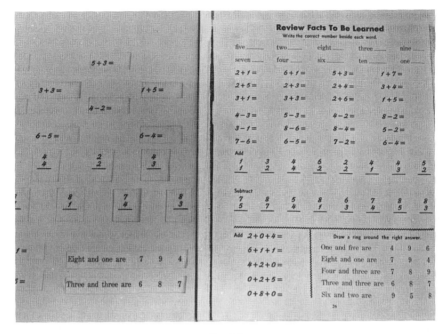

Fig. 6.5

sented by 4̶3, and would be distinguishable from 3̶4. Size cues can also be helpful. For example, 43 could be written

$$4\ 3$$

indicating the increased value of digits to the left. These crutches are meant to be temporary and should gradually be phased out as the children become capable of distinguishing between right and left.

Place-value problems resulting from right-left disorientation often extend to operations on multidigit numbers because these children have difficulty determining where to begin adding or subtracting. It is often helpful to devise a simple marking system that will remind the children how to proceed. For example, a green dot placed above the units column and an arrow pointing left

$$\begin{array}{r} 5\ 6 \\ +3\ 3 \end{array}$$

can serve as a reminder to the children to start with the units on the right and proceed left with the tens.

Memory

Assumption 6

Students have the ability to recall mathematical symbols once they have learned their meanings. Some learning disabled children have memory problems that interfere with their ability to recall the configurations of the various numerals. These students may be able to perform computation or recite facts orally, but they typically experience difficulty writing down answers because they are unable to recall what the numerals look like. Since this disability is one of recall as opposed to recognition, the student's dilemma can be eased by using an aid such as a reference chart consisting of the numerals 1 through 100 in rows of ten. For numerals up to 12, often the numerals on the classroom clock serve as the child's "helper." The youngster will recognize the correct numeral(s) and will be able to use these aids as references for written assignments. Similarly, those students who can visualize individual digits but never remember their correct placement in two-digit numbers could benefit from simple charts containing 10, 20, 30, ...

Whereas some children cannot remember what certain numerals look like, others cannot auditorily recall numeral names. This difficulty can be likened to the embarrassing situation of having inexplicably forgotten a friend's name just as you say, "Joan, I'd like you to meet. . . ." These children will have difficulty reading numbers aloud and doing oral calculations. For them, oral drills of any type will prove frustrating. The teacher should avoid placing these students in positions where they are forced to give verbal answers. If oral drills *are* used, these children can either be allowed to write the answers (on the blackboard, on file cards, on magic slates, etc.) or be given multiple-choice options.

The suggestions above describe stop-gap measures that can allow students to complete assignments by *circumventing* the difficulty. The teacher can also provide exercises that help the children *overcome* their memory deficits. For example, the teacher can have these pupils say the numerals aloud (quietly!) while tracing or copying them. Such practice would help stabilize the visual image so that when the numeral name was spoken, it would trigger the visual equivalent. Similarly, saying the numerals while looking at counting sequences would help stabilize the auditory image.

Working from partial to total recall will also increase the children's ability to recall numerals visually or auditorily (Johnson and Myklebust 1967). Revisualization can be accomplished by providing diminishing degrees of cues for the formation of numerals, such as

$$2+3=\text{\small 5} \quad \text{to} \quad 1+4=\text{\small 5} \quad \text{to} \quad 3+2=\text{\small 5}$$

Similarly, decreasing phonemic cues such as "fi-" and then "f-" could serve to trigger auditorially the word *five*.

Since recall improves with rehearsal regardless of the type of memory problem, it is important to expose the children continually to the symbols (visual or auditory) that they have difficulty recalling. However, this should initially be done in a nonthreatening environment where the children are not forced to perform before their classmates.

Assumption 7

Students can recall the proper sequence of steps required in an algorithm once they understand the process and can perform each individual step. The operations of addition and subtraction on single-digit numbers may be acquired easily, but when the operations are applied to multidigit numbers, confusion may ensue. Typically the students will start to add or subtract from the left side of the problem or forget to regroup in appropriate instances. Also, the digits in the minuend of a subtraction problem may be deducted from the subtrahend. These problems may not stem from a lack of understanding of the operation per se but may instead be the result of a memory problem.

Again, as with all memory problems, practice is most beneficial. However, unless the students are practicing the correct procedures, practice is useless, if not detrimental. For this reason it is important that the children initially be provided with cues that will ensure correct performance. The green dot and arrow over the right digit (discussed in assumption 5) will remind the children to start adding or subtracting from the right. Boxes above the top digits will remind the students to determine if regrouping is necessary. If regrouping is required, the students record the notation:

If no regrouping is necessary, the students put an X in the box to indicate that no regrouping is required:

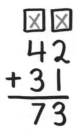

For those children who exhibit inconsistency in subtracting the subtrahend from the minuend, visual cues may again be beneficial. When they understand that the smaller number is to be subtracted from the larger number, perceptual size cues can remind them that the larger number is on top. For example,

$$\begin{array}{r} 63 \\ -48 \\ \hline \end{array}$$

indicates that 48 must be subtracted from 63, and therefore regrouping is required before the subtraction in the right-hand column can be done.

Assumption 8

Students have the ability to memorize basic facts for addition and subtraction once they understand the operational processes. Difficulty recalling the basic facts is not limited to learning disabled children; memorizing the facts is a laborious process for many children. However, for those children with accompanying memory problems, this task may be overwhelming.

One method of helping the child recall these facts is through the use of patterns that build on the child's strengths. Typically the doubles (4 + 4, 5 + 5, 6 + 6, etc.) are among the easiest addition facts for children to recall, whereas many of the addition facts for sums between 10 and 20 are the most difficult. Hence, doubles can be used as an aid in computing some of the more difficult facts through the introduction of the doubles-plus-one and sharing-numbers concepts (Thornton 1977). In the doubles-plus-one strategy, the child is shown the relationship between such facts as $7 + 7 = 14$ and $7 + 8 = 15$: since 8 is one more than 7, $7 + 8$ is one more than $7 + 7$. The sharing-numbers strategy applies to facts where the addends differ by two (e.g., $7 + 9$, $6 + 8$). The student is shown that by subtracting one from the larger addend and adding it to the smaller addend, a familiar double is obtained. In these strategies, the child's strengths (the understanding of the concept of addition and the recall of the sums of doubles) are used to help compensate for a weakness (the memorization of addition facts). Children can also be encouraged to discover and formulate patterns for themselves, since these tend to be more meaningful than externally imposed memorization strategies.

Much drill and practice is necessary for the learning disabled student to gain proficiency in recalling basic facts. Adaptations of the game of Concentration in which the pupils match pairs of number sums that are equal can aid children in remembering the different facts. As the youngsters become involved in the game, they often unknowingly spend time rehearsing the basic addition facts: "I hope I get a $6 + 6$ or a $5 + 7$ or an $8 + 4$." Such a game could be adapted for subtraction facts also.

Drill activities with flash cards can also provide much needed practice. Having the children make their own cards can be a valuable activity in itself if the pupils are not allowed to copy the answers but are encouraged to use concrete materials to verify them. The children should write the problem

$$\begin{array}{r} 1\overset{2}{} \\ -\ 9 \\ \hline \end{array}$$

on one side of the card, and draw a picture of the problem with the answer on the other

ⵘⵘ ⵘⵘ ⵘ ⵘⵘ ⵘ ⵘⵘ ⵘ

3

thus providing a ready reference for the meaning of the process as well as the answer. These cards can then be used for games, such as a variation on War, in which the student with the largest answer wins the trick.

Until pupils gain proficiency in the rapid recall of the basic facts, a number of crutches can be allowed. Addition charts that the children themselves have filled in, finger counting, or hand-held calculators are all measures that can be employed so that the child does not fall behind in learning algorithms that require the memorization of the addition and subtraction facts as a prerequisite.

Symbolic Comprehension

Assumption 9

Students can easily assimilate the mathematical terminology associated with computation skills. The acquisition of new vocabulary can be very difficult for some learning disabled children. For these students, a simple introduction and explanation of terms is usually not sufficient. Before such youngsters can comprehend and use the terms, they must understand the underlying concepts that the words represent. Although it is important for the pupil eventually to use mathematical terms, the concepts should be initially developed without the use of precise terminology. Not until concepts have been intuitively grasped will the children have a base of referents for new vocabulary.

Concepts should be fostered at the concrete level by having the children use manipulative materials. Teachers should introduce the children to the concepts by using *their* language (e.g., "six and one more make seven" rather than "six plus one equals seven"). After the children have spent some time working with the materials, they should be asked to explain what they have done. Actually, this serves a dual purpose: it helps the teacher evaluate the children's comprehension of the concept as well as the exact language they have available. Once the teacher is convinced that the children understand the concept, as evidenced through their actions and language, the mathematically precise terms can be introduced gradually (Scheffelin and Seltzer 1974). This must be done skillfully by repeatedly emphasizing the relationship between the child's terms and the new terms. Understanding the concepts and constant repetition of the precise mathematical terms are the keys to success.

Assumption 10

Students can easily associate symbols with the words and ideas they represent. The symbols used in mathematics can be extremely difficult for some children to cope with. They just cannot associate the symbols with the words or ideas they represent (e.g., the symbol 5 with the spoken or written word five or with the quantity of fiveness; the symbol + with the meaning of addition). As with memory difficulties, symbol association problems are often evidenced in children who can do written computation such as 5 + 2 = 7 but cannot respond correctly to the question, "What does five plus two equal?" Such a problem may arise when a child has learned a rote procedure for determining sums (e.g., moving beads on an abacus) but does not associate what is being done with the oral form of the problem. Unlike children with memory problems, however, these children are unable to answer correctly even if they are allowed to write the answer or are given several answers from which to choose. This is because the association between the written form

and the oral form "five plus two equals seven" was not simply forgotten— it was never made in the first place! Furthermore, when requested to "read"

$$+\ \frac{5}{2}\ \overline{7}$$

(a task that again would require an association between the written and the oral symbolic forms), the child may respond simply by saying "five, two, seven."

Although the causes of this type of difficulty may differ from the causes of memory problems, the corrective procedures are similar. The main objective is to help the children establish associations between equivalent symbols. This can be accomplished by presentations of the *Sesame Street* type, in which a particular symbol is repeatedly shown, pronounced, and spelled out while its meaning is demonstrated in a variety of different situations. Questions requiring the pupils to recognize these connections could then be asked. For example, the teacher could point to each of a variety of symbols (+, =, −, <) asking, "Is this an equal sign?" Similarly, using the same assortment of symbols, the child could be requested to "point to the equals sign."

The pupils could also strengthen their symbol associations through activities involving the assembling of puzzle cards. Such activities would be self-checking because the children would know the symbols were equivalent if the appropriate puzzle pieces fitted together. More challenging variations of this activity could be introduced once the pupils became proficient in assembling the individual pieces. For example, the children could be asked to sequence the necessary cards to read "five plus four equals nine" (see fig. 6.6).

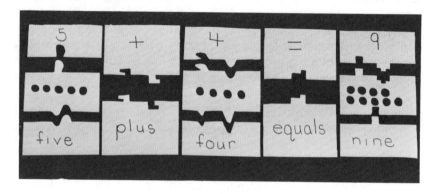

Fig. 6.6

Since the important aspect of remediating this type of difficulty is the simultaneous exposure to equivalent symbols, it is beneficial to give the child a great deal of practice reading examples aloud (without having to work them out) and writing them while a teacher or peer dictates them. Activities like these play a major role in helping the child establish the correct associations between auditory and visual symbols.

Conclusion

Elementary school teachers have traditionally been challenged to find ways of teaching computation to children with diverse aptitudes, skills, and interests. Because of the trend toward mainstreaming in elementary education, this challenge is increasing rather than diminishing. Teachers must now understand and instruct children who formerly would have spent their entire school day in special education classrooms. Some assumptions that teachers traditionally make regarding their pupils' underlying learning processes have been presented. It was suggested that these assumptions be carefully examined when working with learning disabled children.

Practical ideas have been offered to help the classroom teacher meet the needs of these children. The stress has been on techniques that are not "far out" but are adaptations of methods and materials that are commonly used in the regular classroom. The key to the application of these techniques is determining with whom they would be effective. Again it should be emphasized that no learning disabled child will exhibit all the disabilities described in this article, nor will any two learning disabled children have the same combination of disabilities. Therefore, as with all children, the teacher must treat each learning disabled child as a unique person with unique needs. Hence, it is important to realize that these suggestions are not panaceas for *all* learning disabled children.

Further, teachers should recognize that some children have problems so severe or all-pervasive that they need more help than the regular classroom teacher can provide. These children will require more intensive instruction by a learning disabilities specialist.

In summary, it should be acknowledged that good instruction is invaluable for the learning disabled child as well as for the normal child. In other words, the teacher who can analyze a computational skill in terms of its component skills, determine the pupil's proficiencies with regard to those skills, and then present instruction geared to developing the appropriate skill will be a great asset to the learning disabled child. However, such teachers may find that they must go beyond the usual good teaching principles (e.g., instruction should progress from concrete to abstract) and examine the underlying learning processes that the child brings to the instructional situation. In so doing, the teacher may find out why, for the learning disabled child, there is more to addition and subtraction than computation.

BIBLIOGRAPHY

Bartel, Nettie R. "Problems in Arithmetic Achievement." In *Teaching Children with Learning and Behavior Problems,* edited by Donald Hammill and Nettie Bartel. Boston: Allyn & Bacon, 1975.

Green, Roberta. "A Color-coded Method of Teaching Basic Arithmetic Concepts and Procedures." *Arithmetic Teacher* 17 (March 1970): 231–33.

Johnson, Doris J., and Helmer R. Myklebust. *Learning Disabilities: Educational Principles and Practices.* New York: Grune & Stratton, 1967.

Kaliski, Lotte. "Arithmetic and the Brain-injured Child." *Arithmetic Teacher* 9 (May 1962): 245–51.

May, Lola June. *Teaching Mathematics in the Elementary School.* New York: Free Press, 1974.

Moyer, Margaret B. "Mathematics Preparation for Learning Disabilities Teachers." Paper presented at the Annual Spring Meeting of the Wisconsin Mathematics Council, Green Lake, Wisc., May 1975.

National Advisory Committee on Handicapped Children. *Special Education for Handicapped Children.* First Annual Report. Washington, D.C.: U.S. Department of Health, Education, and Welfare, 1968.

Scheffelin, Margaret A., and Carl Seltzer. "Math Manipulatives for Learning Disabilities." *Academic Therapy* 9 (Spring 1974): 357–62.

Thornton, Carol A. "Helping the Special Child Measure Up in Basic Fact Skills." *Teaching Exceptional Children* 9 (Winter 1977): 54–55.

Vygotsky, Lev S. *Thought and Language.* Cambridge, Mass.: The M.I.T. Press, 1962.

7

Teaching Multiplication and Division Algorithms

Donald W. Hazekamp

T HE understanding of multiplication and division is initially developed with small numbers so that concrete materials and pictures can be used to promote pupils' understanding. Such work is a necessary stage in order to have them eventually reach the point of being able to work such examples as 42 × 76 and 24 $\overline{)628}$. Since manipulating objects and interpreting pictures become cumbersome as the numbers become larger, algorithms are used to make computation more efficient. In teaching algorithms, the teacher must build an instructional sequence that fits together underlying principles and concepts.

This article focuses on developing instructional sequences for the conventional two-digit multiplication algorithm and for one- and two-digit divisors in the division algorithm. Base (or grouping) representations for numbers are used with informal reasoning patterns. Base representation means a form that directly indicates the number of each group. Thus, 36 is thought of as "3 tens and 6 ones" and 120 as "12 tens" as well as "1 hundred and 2 tens." In a base model, the total number of units in a number is always present and can be seen. In the example shown in figure 7.1 with base-ten blocks representing 24, pupils see the two tens blocks, and they also see the twenty units.

In contrast, a place-value model such as an abacus could be used (fig. 7.2). One object represents 10, but the total number of units is not seen. Place value, a positional scheme for writing numerals, develops

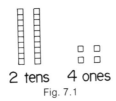

2 tens 4 ones

Fig. 7.1

from experiences with grouping. When numerals are renamed, it is contended that word names (e.g., 2 tens 4 ones) are more closely related to base ideas and expanded notation (e.g., 2 × 10 + 4) is more closely related to place-value ideas. It is evident that base ideas and place-value ideas are not the same (Scrivens 1968).

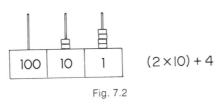

Fig. 7.2

It can be argued that much developmental work with base ideas is needed and that a greater understanding of algorithms may result from stressing base ideas (Scrivens 1968; Smith 1973). Payne and Rathmell (1975) have indicated how thinking in terms of a base representation is essential for algorithmic work. Such thinking can be seen in these situations:

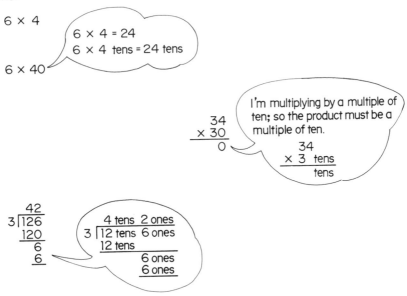

In one study, a base-oriented approach to teaching two-digit multiplication was compared with an approach emphasizing place value and properties (Hazekamp 1976). In the base-oriented approach, both word-name numerals and base models were used in the developmental phase. Thus, 3×40 was thought of like this:

3×4 tens $= 12$ tens
12 tens $= 120$
so $3 \times 40 = 120$

3 groups of 4 tens

The place-value approach involved expanded notation and the associative property of multiplication without specific reference to the physical grouping of ten:

$$3 \times 40 = 3 \times (4 \times 10)$$
$$= (3 \times 4) \times 10$$
$$= 12 \times 10$$
$$= 120$$

The base-oriented approach was found to be more effective than the place-value approach. Therefore, the instructional sequence that follows uses a base-oriented approach.

The Multiplication Algorithm

In this section, an instructional sequence leading to teaching the algorithm for two two-digit factors is presented. The sequence has been used successfully with fourth- and fifth-grade pupils. Since problems that arise at more difficult levels are often related to pupils' insight and performance at lower stages, the development of prerequisite concepts and skills is considered first, followed by ideas on special two-digit products. Finally, procedures for teaching the two-digit-by-two-digit algorithm are discussed.

Developing prerequisite concepts and skills

Knowledge of basic multiplication facts

Pupils should be able to give the products for all basic multiplication facts that are to be used in the two-digit work. Early work on the algorithm should focus on the new ideas being presented, not on difficulties with the multiplication facts. (It may be necessary, however, for some children to have charts of the basic facts or a calculator available for reference.)

The structure of the base-ten numeration system and renaming skills

Pupils should understand the structure of the base-ten numeration system—particularly the 10-to-1 and 1-to-10 relationships, which are needed in renaming numbers—and they should understand the relationship between base (grouping) ideas and place value as used in written numerals. Many difficulties arise in algorithmic work because numeration ideas are not sufficiently well known for pupils to make a meaningful application of those ideas. Reviewing numeration ideas before and during the study of any of the algorithms is worthwhile.

In order for pupils to develop a *thinking model* that will be useful in their work with algorithms, it is essential that they have experiences with base models. Some effective ways for them to develop this thinking model are to do grouping activities, give different names for numbers, and draw appropriate pictures (fig. 7.3).

The grouping-by-tens pattern can be demonstrated by using a set of base-ten blocks or by using ones squares, tens strips, and hundreds squares made from centimeter graph paper.

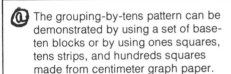

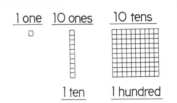

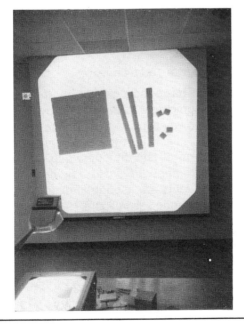

When you are working with a group of pupils, the blocks or other objects can be put on the stage of an overhead projector and projected on the screen.

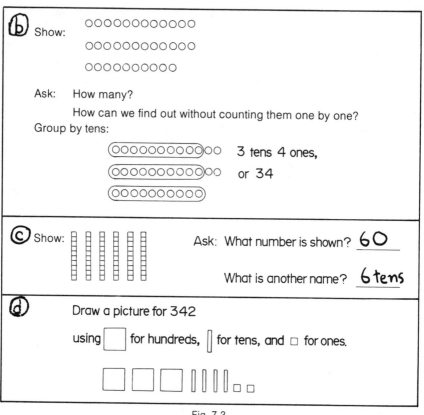

Fig. 7.3

Multiplying multiples of 10 and multiples of 100 by ones

Teaching multiplication beyond the basic facts cannot be done effectively or efficiently if pupils are not competent in multiplying by tens, by hundreds, and later by thousands. The key aspects at this stage are to relate products such as 4 × 30 to basic facts and to have pupils see the pattern. For example, in the multiplying of tens by ones, pupils need to see the pattern *number of ones times number of tens is a number of tens* and see how the basic multiplication facts are used in finding the product. Pupils need to establish a thinking model, as in figure 7.4.

$3 \times 70 = \square$

Fig. 7.4

This can be done by engaging them in activities like those that follow.
▶ Initially, have pupils compare two arrays and find the products.

4 × 3 ones   4 × 3 tens

4 × 3 = 12   4 × 3 tens = 12 tens; so 4 × 30 = 120

▶ Discuss them by asking questions such as these:
- "How are the two arrays alike?" (Both have four rows with three in each row.)
- "How are they different?" (The first has ones in each row; the second has tens.)
- "How are their products different?" (In the first, we get twelve ones; in the second, twelve tens.)

▶ Then have pupils work examples in vertical form:

$$\begin{array}{r} 6 \\ \times\ 3 \\ \hline 18 \end{array} \qquad \begin{array}{r} 6\ \text{tens} \\ \times\ 3 \\ \hline 18\ \text{tens} \end{array} \quad \text{so} \quad \begin{array}{r} 60 \\ \times\ 3 \\ \hline 180 \end{array}$$

A similar type of pattern can be shown for multiples of one hundred.

After the concept is introduced, both oral and written practice should be given, making sure that all types of examples are included.

$$\begin{array}{r} 4\ \text{tens} \\ \times 5 \\ \hline 20\ \text{tens} \end{array} \quad \text{so} \quad \begin{array}{r} 40 \\ \times\ 5 \\ \hline 200 \end{array} \qquad \begin{array}{r} 5\ \text{hundreds} \\ \times\ 6 \\ \hline 30\ \text{hundreds} \end{array} \quad \text{so} \quad \begin{array}{r} 500 \\ \times\ \ 6 \\ \hline 3000 \end{array}$$

The one-digit-by-two-digit multiplication algorithm

Multiplication with two two-digit numbers requires pupils to have competence with the one-digit-by-two-digit algorithm. The stages for this procedure are given briefly to show how and where it fits into the sequence and the way the first three prerequisites, particularly base representations, are used.

Distributivity is the property that underlies the multiplication algorithm. Children's understanding of this property develops slowly, and for many it is not a stable concept until sixth grade or later (Crawford 1965;

Flournoy 1967 [a] and [b]). Thus, in the procedure described here the distributive property is used informally.

▶ Begin by using concrete models. For example, in multiplying 3 × 34, show an array of 3 thirty-fours, and have pupils note that there are 3 thirties and 3 fours.

Pupils should think of this as

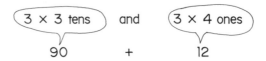

$$\boxed{3 \times 3 \text{ tens}} \quad \text{and} \quad \boxed{3 \times 4 \text{ ones}}$$

$$90 \qquad + \qquad 12$$

Several examples should be worked out using both physical and pictorial models.

▶ Then relate the thinking pattern established in the model to the steps in the algorithm, being careful to note that we multiply twice.

	tens	ones
	3	4
×		3

First, multiply the ones (3 × 4 ones) ⟶ 1   2

Second, multiply the tens (3 × 3 tens) ⟶ 9   0

Third, add (90 + 12) ⟶ 1   0   2

It is crucial that pupils understand that they are multiplying 3 × 4 ones and 3 × 3 tens in order to avoid such incorrect procedures as these:

```
    34                34
  × 3      and      × 3
  ----              ----
  912                12
                      9
                    ----
                     21
```

▶ Move to the short form, for proficiency with it is needed in the two-digit-by-two-digit algorithm. When the short form is taught, attention must be given to the new component, renaming, which is done when the product of the ones is greater than nine. We still multiply twice.

First, multiply ones and <u>RENAME</u>

Second, multiply tens and <u>ADD EXTRA TENS</u>

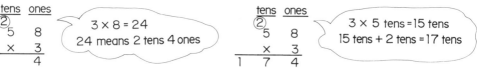

To help pupils recognize when they need to rename, give separate examples that focus on this aspect:

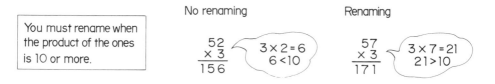

Circle the examples in which you need to rename:

$$\left(\frac{56}{\times 4}\right) \quad \frac{72}{\times 3} \quad \frac{64}{\times 2} \quad \left(\frac{48}{\times 6}\right)$$

▶ To help build the relationship between grouping ideas and place value in written numerals, have pupils record on a chart as examples are shown.

Number	Thousands	Hundreds	Tens	Ones	Word Name
34			3	4	3 tens 4 ones
60			6	0	6 tens 0 ones
342		3	4	2	3 hundreds 4 tens 2 ones

Thinking of numbers as multiples of ten and a hundred is an important skill. Pupils need to recognize that if a number is a multiple of ten, then there will be no ones, which is shown by a zero in the ones place (or by zeros in the tens and ones places if the number is a multiple of a hundred). This type of thinking is needed for understanding when a multiple of ten (or a hundred) is used as a factor; pupils discover that the product must also be a multiple of ten (or a hundred).

$$5 \times 60 = 30 \text{ tens} = 300 \qquad 30 \times 24 = 72 \text{ tens} = 720$$

After a little work, pupils can readily state that 60 = 6 tens, but they are not as confident when numbers are larger. They do not readily think of 230 as 23 tens, 2400 as 24 hundreds, or 30 tens as 3 hundreds. The tens strips and hundreds squares can be used to develop these concepts. Begin by showing pupils examples and having them answer questions (fig. 7.5).

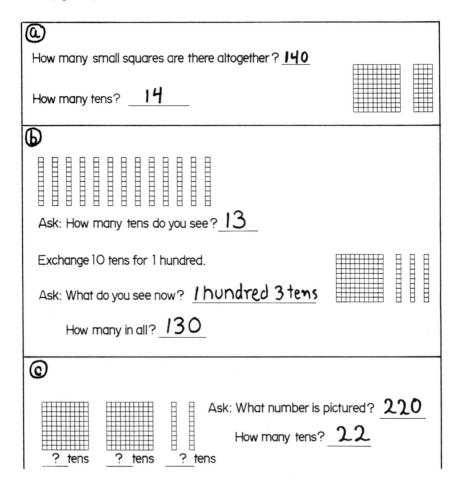

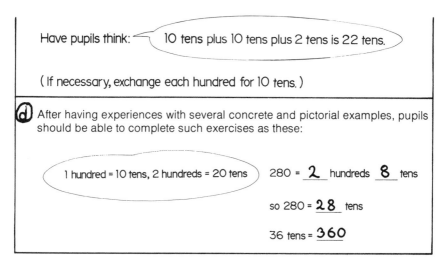

Have pupils think: 10 tens plus 10 tens plus 2 tens is 22 tens.

(If necessary, exchange each hundred for 10 tens.)

ⓓ After having experiences with several concrete and pictorial examples, pupils should be able to complete such exercises as these:

1 hundred = 10 tens, 2 hundreds = 20 tens   280 = **2** hundreds **8** tens

so 280 = **28** tens

36 tens = **360**

Fig. 7.5

Note that the ability to interpret three-place multiples of ten in two ways (i.e., 120 as 1 hundred 2 tens and also as 12 tens) is used extensively in this section and the work that follows. The corresponding skill of interpreting four-place multiples of a hundred in two ways (i.e., 1200 as 1 thousand 2 hundreds and as 12 hundreds) is also important. Thus, illustrations and examples involving the relationships between ones, tens, hundreds, and thousands should also be used (figs. 7.6, 7.7, and 7.8).

14 hundreds = **1400**

2000 = **20** hundreds

Fig. 7.6

The main objectives of the activities above are to have pupils give different names for the same number and to establish the thinking model. In developing these ideas, teachers should have children use both physical and pictorial models (figs. 7.9 and 7.10). Ask pupils to demonstrate the number with the models as well as to write out the symbols. The amount of time needed for this review—probably two or three hours—will depend on how familiar the children are with base representations and the models used.

Special two-digit products

Examples such as 30 × 60 and 26 × 40 are special types of two-digit-by-two-digit multiplication—they do not require the use of the usual al-

Fig. 7.7. Exchanging 10 hundreds squares for 1 thousands block

Fig. 7.8. Using strips and squares

Fig. 7.9. Using models

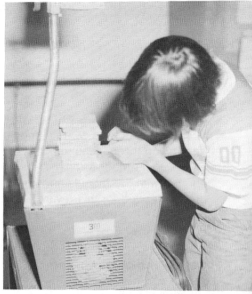

Fig. 7.10. Counting 12 hundreds

gorithm. They should be taught prior to the teaching of the two-digit algorithm itself. Some reasons:

- They can be calculated more efficiently with a one-step procedure than by following all steps in the two-digit algorithm.
- Finding the product of a multiple of ten and a two-digit number is a skill directly applicable to the two-digit-by-two-digit algorithm.
- It has been my experience that pupils learn to recognize these types better when they are taught prior to instruction on the two-digit algorithm rather than after such instruction.

Multiplying tens by tens

In developing numeration ideas, pupils have learned that 10 tens = 1 hundred. From a hundreds square they have seen that 1 ten × 1 ten = 1 hundred. Those ideas can be extended to the multiplying of a multiple of ten by a multiple of ten, where pupils need to see that *a number of tens times a number of tens is a number of hundreds*. This pattern can be established by having pupils see several examples where *m* tens × *n* tens = *(m × n)* hundreds (see fig. 7.11).

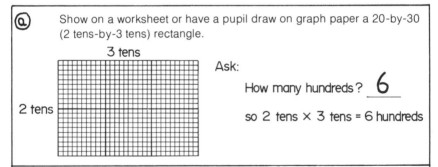

ⓐ Show on a worksheet or have a pupil draw on graph paper a 20-by-30 (2 tens-by-3 tens) rectangle.

3 tens

2 tens

Ask:

How many hundreds? **6**

so 2 tens × 3 tens = 6 hundreds

ⓑ Use an illustration and make a chart like the one below, in which pupils can see the pattern.

"Suppose we have 10 bags of marbles. Each bag has 30 marbles. How many marbles are there altogether?

(Pupils may have to count by 30s.) Record on the chart below.

"Suppose there are 10 more bags of marbles. How many marbles altogether now?"

Continue the process of adding ten more bags and recording results. While filling the chart, emphasize multiplication, for many pupils can find the total by adding 300 without noting the multiplication involved.

Number of bags	Number in each bag	What do we have?	How many altogether?
10	30	10 sets of 30	$10 \times 30 = 300$
20	30	20 sets of 30	$20 \times 30 = 600$
30	30	30 sets of 30	$30 \times 30 = 900$
40	30	40 sets of 30	$40 \times 30 = 1200$

Discuss the chart and rename each number of the number sentences in the last column:

$$1 \text{ ten} \times 3 \text{ tens} = 3 \text{ hundreds}$$
$$2 \text{ tens} \times 3 \text{ tens} = 6 \text{ hundreds}$$
$$3 \text{ tens} \times 3 \text{ tens} = 9 \text{ hundreds}$$

The discussion should lead pupils to see the pattern m tens $\times$ n tens = $(m \times n)$ hundreds and develop a thinking model:

4 tens   8 tens $4 \text{ tens} \times 8 \text{ tens} = 32 \text{ hundreds}$

$$40 \times 80 = \qquad 3200$$

(c) It is particularly important to include examples that involve potential difficulties with zero and place value.

$$40 \times 50 = 4 \text{ tens} \times 5 \text{ tens} = \underline{\qquad} \text{ hundreds} = 2000$$
$$50 \times 60 = 5 \text{ tens} \times 6 \text{ tens} = \underline{\qquad} \text{ hundreds} = 3000$$

(d) The final process might be summarized by this example:

Jack has to multiply 50×70. See how he does it.

$50 \times 70 = \underline{35}$

so $50 \times 70 = \underline{3500}$

FIRST, I FIND THE PRODUCT OF THE TENS DIGITS, $5 \times 7 = 35$. SECOND, I WRITE TWO ZEROS TO SHOW A MULTIPLE OF A HUNDRED.

Fig. 7.11

Finding the product of a power of ten and a two-digit number

Commutativity has been shown to be the least difficult of the whole-number properties (Flournoy 1967 [a] and [b]; Crawford 1965). Yet children who can readily see and use the commutative property with smaller numbers (e.g., 3 × 5 = 5 × 3) are unable to see or use the property with larger numbers. Even after writing 380 as the response for 38 × 10, pupils often give 138 or 1038 as responses for 10 × 38. They do not think of 10 thirty-eights as being the same as 38 tens. The use of arrays is an effective way to demonstrate the concept. For instance, show pupils a 13-by-10 array which shows 13 tens (fig. 7.12). Thus 13 tens = 130, or 13 × 10 = 130.

Now turn the array so that it shows 10 thirteens (fig. 7.13). Thus, we can say that 10 thirteens = 13 tens and that by renaming, 1 ten × 13 = 13 tens, or 10 × 13 = 130. More examples using larger numbers, such as 10 × 24 and 10 × 38, need to be demonstrated too.

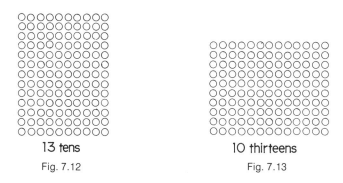

13 tens   10 thirteens

Fig. 7.12   Fig. 7.13

Since the array model is difficult to use when multiplying by one hundred, charts like the one in figure 7.14 are helpful and promote the recognition of patterns.

	Thousands	Hundreds	Tens	Ones
1 × 52			5	2
10 × 52		5	2	0
100 × 52	5	2	0	0

So 1 hundred × 52 = 52 hundreds

Fig. 7.14

Not only is an understanding of this concept important on its own, but it is needed as a prerequisite for multiplying any multiple of ten times a two-digit number.

Multiplying a two-digit number by a multiple of ten

The majority of errors that pupils make in two-digit multiplication are in finding the second partial product in which a two-digit number is multiplied by a multiple of ten. This concept must be developed carefully to give them a clear understanding of the underlying place-value ideas. An effective way to introduce this concept is to discuss with pupils a series of four or five statements that focus on multiplying by multiples of ten. For example:

$$
\begin{array}{rcl}
1 \times 42 &=& 42 \\
1 \text{ ten} \times 42 &=& 42 \text{ tens} \\
\text{so } 10 \times 42 &=& 420
\end{array}
\qquad
\begin{array}{rcl}
2 \times 42 &=& 84 \\
2 \text{ tens} \times 42 &=& 84 \text{ tens} \\
\text{so } 20 \times 42 &=& 840
\end{array}
$$

$$
\begin{array}{rcl}
3 \times 42 &=& 126 \\
3 \text{ tens} \times 42 &=& 126 \text{ tens} \\
\text{so } 30 \times 42 &=& 1260
\end{array}
\qquad
\begin{array}{rcl}
4 \times 42 &=& 168 \\
4 \text{ tens} \times 42 &=& 168 \text{ tens} \\
\text{so } 40 \times 42 &=& 1680
\end{array}
$$

Pupils should see that when one of the factors is a multiple of ten, the product is also a multiple of ten. The beginning exercises should highlight this idea:

(a) $4 \times 14 = 56$ (b) $8 \times 32 = 256$

4 tens $\times 14 = \mathbf{56}$ tens   8 tens $\times 32 = \mathbf{256}$ tens

so $40 \times 14 = \mathbf{560}$ so $80 \times 32 = \mathbf{2560}$

These exercises can be followed by exercises that focus attention on what numbers need to be multiplied to find the total number of tens:

(a) $\begin{array}{r} 32 \\ \times 4 \\ \hline 128 \end{array}$ 4 tens $\times 32 = \mathbf{128}$ tens   (b) $\begin{array}{r} 54 \\ \times 6 \\ \hline 324 \end{array}$ 6 tens $\times 54 = \mathbf{324}$ tens

so $40 \times 32 = \mathbf{1280}$ so $60 \times 54 = \mathbf{3240}$

And finally, exercises in vertical form can be provided:

(a) $\begin{array}{r} 26 \\ \times 3 \\ \hline 78 \end{array}$ $\begin{array}{r} 26 \\ \times 3 \text{ tens} \\ \hline \mathbf{78} \text{ tens} \end{array}$ so $\begin{array}{r} 26 \\ \times 30 \\ \hline \mathbf{780} \end{array}$

(b) $\begin{array}{r} 48 \\ \times 3 \\ \hline 144 \end{array}$ $\begin{array}{r} 48 \\ \times 30 \\ \hline \mathbf{1440} \end{array}$

(c) $\begin{array}{r} 53 \\ \times 40 \\ \hline \mathbf{2120} \end{array}$

In the final vertical form, the following type of thinking is being generated:

What must be done to multiply $\begin{array}{r} 53 \\ \times 40 \end{array}$?

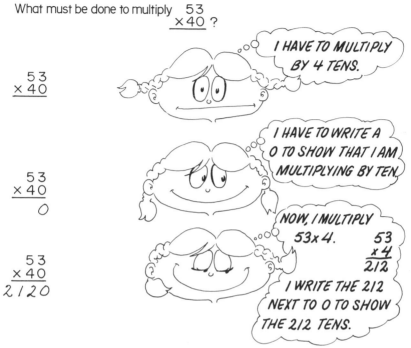

$$\begin{array}{r} 53 \\ \times 40 \\ \hline \end{array}$$

$$\begin{array}{r} 53 \\ \times 40 \\ \hline 0 \end{array}$$

$$\begin{array}{r} 53 \\ \times 40 \\ \hline 2120 \end{array}$$

Since this concept—that when one of the factors is a multiple of ten, the product is also a multiple of ten—is the major new idea needed for the two-digit multiplication algorithm, more time and practice must be given at this stage than at previous stages to develop the level of mastery necessary for its effective use in the algorithm.

The two-digit-by-two-digit algorithm

The approach presented involves fitting together the two steps needed to find the product. Each step has already been presented separately.

Multiply by the ones	Multiply by the tens
$\begin{array}{r} 32 \\ \times 4 \\ \hline \end{array}$	$\begin{array}{r} 32 \\ \times 20 \\ \hline \end{array}$

In learning the algorithm, pupils must see how these component skills learned earlier fit into the whole process. Thus, it is more important in the initial lessons on the algorithm for teachers to discuss the steps in the process and for pupils to give attention to them than for the children to work a large number of examples (see fig. 7.15).

The steps

Say: "You already know how to find the products of 32 × 4 and 32 × 20.
    Now you can find the product of 32 and 24."

Then demonstrate how:

First, multiply 32 by 4.

$$\begin{array}{r} 32 \\ \times\ 4 \\ \hline 128 \end{array} \qquad \begin{array}{r} 32 \\ \times 24 \\ \hline 128 \end{array} \text{(first partial product)}$$

Second, multiply 32 by 20.

$$\begin{array}{r} 32 \\ \times 20 \\ \hline 640 \end{array} \quad \underset{\text{add}}{\text{then}} \quad \begin{array}{r} 640 \\ \hline 768 \end{array} \text{(second partial product)}$$

Have pupils note that they multiply twice in the algorithm—first by the ones
and then by the tens—and then add.

Fig. 7.15

In the beginning exercises, difficult multiplication facts and renaming
should be kept to a minimum. It is helpful to have exercises where each
part is worked separately (fig. 7.16).

Multiply by ones:
$$\begin{array}{r} 42 \\ \times 3 \\ \hline 126 \end{array} \qquad \begin{array}{r} 42 \\ \times 23 \\ \hline 126 \end{array}$$

Multiply by tens:
$$\begin{array}{r} 42 \\ \times 20 \\ \hline 840 \end{array}$$

Then add:
$$\begin{array}{r} 840 \\ \hline 966 \end{array}$$

Fig. 7.16

To focus attention on the second partial product, it is helpful to have
exercises in which the first partial product is supplied and the pupils need
only compute the second one (fig. 7.17). This is also helpful when renaming is introduced in the second partial product.

Exercises should be sequenced carefully so that pupils are not confronted with several types of difficulty at one time. Regrouping in multiplication should not occur in either of the partial products in the initial learning stage. Later, renaming in the first partial product and then in the
second can be introduced.

```
              32
             ×26
Multiply by ones:   192
Multiply by tens:   640
Add:
                   832
```

Fig. 7.17

After pupils have demonstrated that they can do all the steps in the algorithm, the following procedure might be used to eliminate the use of the crutches.

When multiplying by ones in the example shown in figure 7.18, have pupils cover up the 3 tens of 35 so all that shows is

			tens	ones
47			4	7
× 5			× 3	5
	(5 × 47)	2	3	5

Then multiply by ones. Next have them cover up the 5 ones; now all that shows is

(5 × 47)		2	3	5
(30 × 47)	1	4	1	0
Add	1	6	4	5

47
×3

Fig. 7.18

Multiply by tens. When this is done on the chalkboard, using a card with *tens* written on it is helpful when covering up the ones (fig. 7.19).

Fig. 7.19

Even though drill is important, an excessive number of exercises is not necessary. The important aspects appear to be these:

● A clear understanding of the underlying place-value ideas
● A careful development of each stage, with emphasis on the component concepts
● The development of mastery at each level before moving to the next

The Division Algorithm

For many teachers and pupils, long division is the most difficult of the computational algorithms. Pupils need an understanding of, and proficiency with, such subskills as numeration ideas, subtraction, and multiplication, and they need an understanding of how these skills fit together in the division algorithm. Over forty years ago, Knight (1930, p. 162) stated: "Learning division is a problem of educational methods as well as a problem of mathematics." The sequence for teaching division must be a carefully developed series of steps that tries to solve both these problems.

The subtractive method found in Isaac Greenwood's *Arithmeticks* of 1729 and the distributive method used in a text by Calandri in 1491 are still being used (fig. 7.20).

Subtractive

$$24\overline{\smash)753}$$
$$-480 \quad \text{20 (twenty-fours)}$$
$$\overline{273}$$
$$-240 \quad \text{10 (twenty-fours)}$$
$$\overline{33}$$
$$-\underline{24} \quad \text{1 (twenty-four)}$$
$$9$$

Answer: 31 r 9

Distributive

$$\begin{array}{r} 31\;r\;9 \\ 24\overline{\smash)753} \\ -720 \\ \overline{33} \\ -\;24 \\ \overline{9} \end{array}$$

Fig. 7.20

The instructional sequence to be described will lead to the development of the distributive division algorithm. In some examples, a subtractive form is used so that all aspects of the work can be noted. As in the development of multiplication algorithms, grouping representations for numbers and informal reasoning patterns will be employed. First to be considered will be prerequisite concepts and skills, followed by the development of the division algorithm with one-digit divisors and two-digit quotients. Brief discussions of work with larger quotients and finally with larger divisors conclude the section.

Developing prerequisite concepts and skills

Three ideas are crucial aspects of division and should be carefully developed prior to any work with the algorithm.

Relating the idea of division to multiplication

First, pupils need to see the relationship between multiplication and division. (See fig. 7.21.)

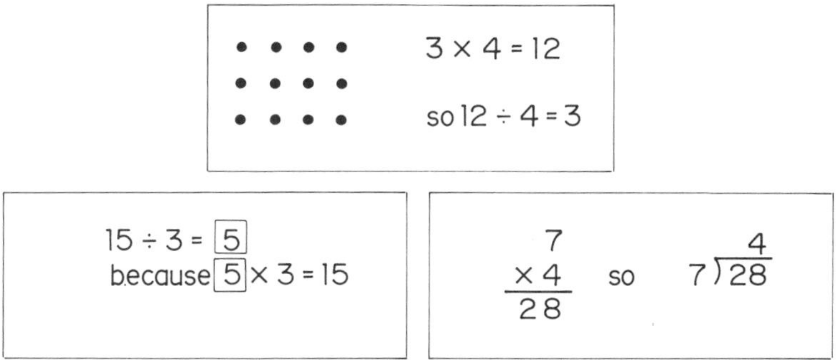

$$3 \times 4 = 12$$
$$\text{so } 12 \div 4 = 3$$

$$15 \div 3 = \boxed{5}$$
$$\text{because } \boxed{5} \times 3 = 15$$

$$\begin{array}{r} 7 \\ \times 4 \\ \hline 28 \end{array} \quad \text{so} \quad 7\overline{)28}^{\,4}$$

Fig. 7.21

To see this, they need to work with missing-factor statements together with their equivalent division statements:

$$\Delta \times 6 = 24 \text{ is equivalent to } 24 \div 6 = \Delta$$

This will help pupils to solve simple division examples using multiplication facts, as in figure 7.22.

Interpreting division as finding a quotient and remainder

Pupils need to have proficiency with simple division with remainders before proceeding to work with the division algorithm (see, for example, fig. 7.23). The concept that in any division example the largest possible quotient must be found needs to be taught early when appropriate concrete models can be used.

Fig. 7.22

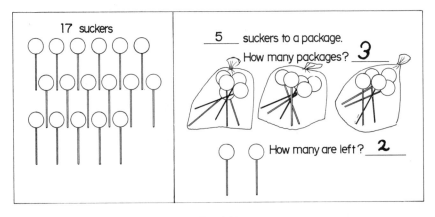

Fig. 7.23

Furthermore, these situations need to be related to multiplication and to the simple algorithmic form:

$$17 = (\boxed{3} \times 5) + \triangle{2}$$
$$\text{or } 17 = \boxed{3} \text{ fives} + \triangle{2}$$

$$\begin{array}{r} 3\,r2 \\ 5\overline{)17} \\ \underline{15} \\ 2 \end{array}$$

These types of problems help pupils to think about what a division example means and what is being asked:

The two types of physical interpretations of division

Pupils need to see division in real-world settings. That is, they need to work with both partition and measurement situations. A partition problem asks such questions as "How many are there in each group?" when the total and the number of groups are known:

Jack has 6 cookies.

How many will he and his 2
friends get if all 3 children get
the same number?

A measurement problem asks, "How many groups are there?" when the total and the number in each group is known:

There are 12 cookies.

Each person gets 3 cookies.
How many will get cookies?

An understanding of partition ideas is particularly helpful in the development of the distributive division algorithm. Partition problems are more difficult for children than measurement problems. They have difficulty recognizing that they must divide in partition situations. It is helpful for them to partition sets of objects in order to gain understanding (see fig. 7.24).

12 blocks

Make 3 equal sets by placing the blocks in each circle until all the blocks are distributed.

Now how many blocks are in each set?

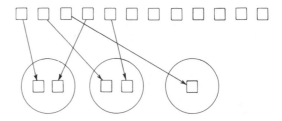

Fig. 7.24

Situations should also be included where there will be blocks left over.

It is important for children to associate the division operation with partition situations. It is not important that they be able to name whether the situation is measurement or partition division, but they should experience such situations so they can see that each is related to division.

Developing the division algorithm: one-digit divisors, two-digit quotients

It is most important that initial work with the division algorithm be slowly and carefully developed. If pupils learn to divide confidently and competently at this level, the later work will move far more smoothly.

Algorithm subskills

Work on the division algorithm begins with the development of three subskills of the algorithm. A brief presentation of each follows.

Dividing multiples of ten by ones. The idea to be developed is this: *A number of tens divided by a number of ones is a number of tens.* The idea can be related to multiplication and to partition division by having pupils think about and describe pictorial models like the one in figure 7.25, both in terms of multiplication and division.

Multiplication	Division
3 sets, 5 tens in each set	15 tens all together
3 × 5 tens = 15 tens	Make 3 sets.
So 3 × 50 = 150 altogether.	15 tens ÷ 3 = 5 tens
	So 150 ÷ 3 = 50 in each set.

Fig. 7.25

In this illustration the grouping thinking model is being used, and the need for the ability to partition is highlighted. After the concept has been established, practice will be needed with exercises like these:

$24 \div 8 = 3$

24 tens $\div 8 = \underline{3}$ tens

so $240 \div 8 = \underline{30}$

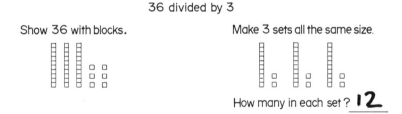

36 tens $\div 9 = 4$ tens

$360 \div 9 = 40$

Partitioning tens and ones. The pupils' earlier skill of partitioning sets into equivalent subsets can now be applied to dividing sets that have both tens and ones. See figure 7.26.

36 divided by 3

Show 36 with blocks.   Make 3 sets all the same size.

How many in each set? **12**

Fig. 7.26

This type of demonstration shows how place-value ideas are involved. Pupils should also see (or you should point out to them) that they need to divide more than once. For this example, they need to divide twice: first, divide the tens, and second, divide the ones. This procedure can be demonstrated with blocks.

$128 \div 4 =$

Show 128 as 12 tens and 8 ones.

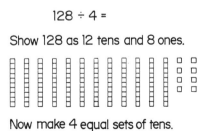

Now make 4 equal sets of tens.

Think: Make 4 equal sets of tens. How many in each set?

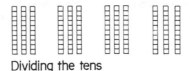

Dividing the tens

Then make 4 equal sets of ones.

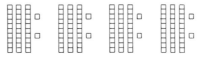

Dividing the ones

Now put the pieces together to get 4 equal sets. So 12 tens 8 ones ÷ 4 = 3 tens 2 ones, or 128 ÷ 4 = 32.

After you have demonstrated the concept with the blocks, it is helpful to have pupils work examples by drawing pictures of what happens using

▯ for tens and ▪ for ones.

The crucial part of developing the idea of the need to divide twice is getting pupils to think of the dividend as groups of tens and ones so they can divide the tens and then the ones.

186 ÷ 3 =

18 TENS 6 ONES.
MAKE 3 SETS.

18 tens 6 ones ÷ 3 = 6 tens 2 ones;
so 186 ÷ 3 = 62

Finally, the thinking pattern needs to be related to the conventional notation: note again that it is necessary to divide twice.

$$\frac{\text{? tens ? ones}}{3\;)\overline{24\text{ tens }9\text{ ones}}} \qquad \text{so} \qquad \frac{\text{?}\quad\text{?}}{3\;)\overline{2\quad4\quad9}}$$

Finding the tens in the quotient. The idea of first dividing the tens has been established. Now the focus is on developing a thinking pattern to find the number of tens in the quotient. For example, when dividing 72 by 3, have the pupils think:

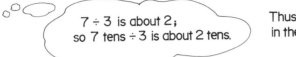

7 ÷ 3 is about 2;
so 7 tens ÷ 3 is about 2 tens.

Thus the quotient is in the 20s.

To gain skill, pupils should work such examples as these:

(a) 26 ÷ 6 is about **4**   (b) 16 tens ÷ 5 is about **3** tens

So 26 tens ÷ 6 is about **4** tens.

Also have them work examples in which they need to find only the tens:

$$\begin{array}{r} 2\,0 \\ 4\overline{)92} \end{array} \qquad \begin{array}{r} 8\,0 \\ 3\overline{)252} \end{array}$$

Teaching the algorithm

The three skills described in the preceding section are component parts of the division algorithm. Now we shall focus on putting these parts together and on learning the steps in the process. The preceding work was done so that pupils will be able to understand what is going on in each step.

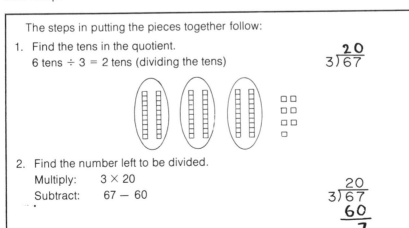

The steps in putting the pieces together follow:

1. Find the tens in the quotient.
    6 tens ÷ 3 = 2 tens (dividing the tens)

    $$\begin{array}{r} 2\,0 \\ 3\overline{)67} \end{array}$$

2. Find the number left to be divided.
    Multiply: 3 × 20
    Subtract: 67 − 60

    $$\begin{array}{r} 20 \\ 3\overline{)67} \\ \underline{60} \\ 7 \end{array}$$

3. Find the ones in the quotient.

   7 ÷ 3 is about 2. (dividing the ones)

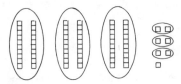

4. Find the remainder.

   Multiply: 3 × 2

   Subtract: 7 − 6

5. Find and write the quotient.

   20 + 2 = 22

   plus the remainder

   The quotient is 22, and the remainder is 1.

At first pupils may forget that they must divide twice; they will have to be reminded to divide the ones that are left. Others feel that they are finished when they have found the tens in the quotient:

$$
\begin{array}{r}
20\ \text{r}11 \\
3\,\overline{)\,71} \\
60 \\
\hline
11
\end{array}
$$

The use of models may help pupils see that there are ones left that still need to be divided (fig. 7.27).

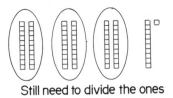

Still need to divide the ones

Fig. 7.27

As children learn the steps, partially worked examples like the following help them focus on particular steps and help ensure that they work the examples correctly from the beginning.

Find the quotient and the remainder:

```
      38 r 2                      72 r 2
        8                          2
       30                         70
    4)154                     4)290
     -120   4 × 30            -280   4 × 70
       34                       10
      -32   4 × 8              -8   4 × 2
        2                        2
```

Copy and complete:

```
        r                          _____
       40
    8)362                      3)227
     -320   8 × 40            ___   3 × 70
       42
      ___   8 × ___          ___   3 × 5
```

It will be necessary to review the steps each day for the first few days. After pupils have a command of the steps and can work examples by themselves, they can be shown how to check their work.

```
    78 r 3     Think:                                   Check:
      8                                                    78
     70          Does 315 = (78 × 4) + 3?                 × 4
  4)315                                                   312
    280                                                   + 3
     35                                                   315
     32
      3
```

The checking process is important, since it reinforces the meaning of division and what the quotient represents.

Moving to the short form

Work can be started with the short form after pupils understand the steps of the longer form and gain confidence in working with it. Emphasize the steps and new parts while showing them the new form.

Tens Ones
6 3 r2
4$\overline{)254}$
-240 4 × 60
1 4
-12 4 × 3
2

Steps

1. Divide to find the tens in the quotient. *Write 6 in the tens place.*

2. Divide to find the ones in the quotient. *Write 3 in the ones place.*

3. Show the remainder. *Write r 2.*

One-digit divisors: quotients "in the hundreds"

This work is an extension of the previous work with the following idea added: *A number of hundreds divided by a number of ones is a number of hundreds.* This idea can be related to the previously learned skills of dividing tens by ones and of multiplication by providing examples like these:

$8 \div 4 = 2$

$8 \text{ tens} \div 4 = \underline{2} \text{ tens}$

$8 \text{ hundreds} \div 4 = \underline{2} \text{ hundreds}$

$35 \div 5 = \underline{7}$

$350 \div 5 = \underline{70}$

$3500 \div 5 = \underline{700}$

Directions for the algorithm can be outlined in a series of four steps.

1. Divide the *hundreds* first.

3$\overline{)1573}$

Write the hundreds in the quotient.

500
3$\overline{)1573}$

3. Find the number of tens and ones (as you have done before)

2. Find the number left to be divided.

500
3$\overline{)1573}$
-1500 500 × 3
73

524 r1
4
20
500
3$\overline{)1573}$
−1500
73
-60 20 × 3
13
12 4 × 3
1

4. Find the quotient and remainder.

Particular attention needs to be given to examples that have zeros in the quotient, such as:

$$\begin{array}{r} 204 \\ 6\,)\overline{1224} \end{array} \qquad \begin{array}{r} 240\ r\ 3 \\ 6\,)\overline{1443} \end{array}$$

It is not uncommon for pupils to get 24 for the answer to these particular examples. Their attention needs to be focused on this problem, so that a thinking pattern such as the following is attained:

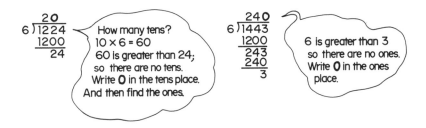

Pupils need to be able to identify the "range" of the quotient; that is, to be able to tell whether the quotient is less than 10, between 10 and 100, or greater than 100. They need to be able to tell the number of places in the quotient. This can best be done by giving them exercises in which finding the range is the only task. For example:

$$4\,)\overline{53}$$

10 fours is 40
53 is greater than 40
So the quotient is greater than 10.

$$4\,)\overline{238}$$

100 fours is 400
238 is less than 400
So the quotient is less than 100.

Tell which quotients are less than ten, in the tens, or in the hundreds.

$1 \times 5 = 5$	$5\,)\overline{324}$	$5\,)\overline{43}$	$5\,)\overline{837}$
$10 \times 5 = 50$	(tens)	($<$ ten)	(hundreds)
$100 \times 5 = 500$			

The technique can be extended to examples that have larger divisors:

$$1 \times 34 = 34 \qquad 34\overline{)1723}$$
$$10 \times 34 = 340$$
$$100 \times 34 = 3400$$

Are there as many as ten 34s in 1723? **Yes**
Are there as many as 1 hundred 34s
in 1723? **No**
So the quotient is between 10 and 100.

The benefits of this kind of work are twofold:

- It helps pupils to know where to put the first quotient digit by letting them know if they are finding ones, tens, or hundreds.
- It is a generalizable technique that pupils can use through all their division work.

Two-digit divisors

The importance of an understanding of the underlying place-value ideas can be seen very clearly when working with two-digit divisors. Work with two-digit divisors begins with divisors that are multiples of 10. In the preliminary work, pupils should work with such examples as 24 tens ÷ 6 tens = 4 ones and 18 hundreds ÷ 9 tens = 2 tens to help them to develop a thinking pattern for such examples as these:

Both these patterns can be related to multiplication and to the place-value–related patterns of *tens divided by tens are ones* and *hundreds divided by tens are tens.*

$$4 \times 2 \text{ tens} = 8 \text{ tens} \qquad 3 \text{ tens} \times 5 \text{ tens} = 15 \text{ hundreds}$$
$$8 \text{ tens} \div 2 \text{ tens} = 4 \text{ ones} \qquad 15 \text{ hundreds} \div 5 \text{ tens} = 3 \text{ tens}$$
$$\text{so } 80 \div 20 = 4 \qquad \text{so } 1500 \div 50 = 30$$

Place-value ideas can also be seen in approximation, which division with two-digit divisors requires. Number-line models work well for reviewing this concept:

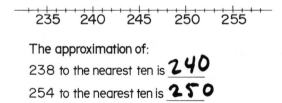

The approximation of:
238 to the nearest ten is **240**
254 to the nearest ten is **250**

In teaching how to find the first digit in the quotient, the teacher has a choice of procedures. It could be found by using approximation or by looking at the range. Figure 7.28 illustrates how approximation can be used together with finding the range.

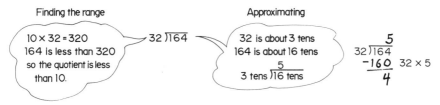

Finding the range

10 × 32 = 320
164 is less than 320
so the quotient is less
than 10.

32)164

Approximating

32 is about 3 tens
164 is about 16 tens

5
3 tens)16 tens

$$\begin{array}{r} 5 \\ 32\overline{)164} \\ -160 \\ \hline 4 \end{array}$$ 32 × 5

Fig. 7.28.

Here, finding the range tells that the quotient is in the ones. Pupils noting this, and knowing the *tens divided by tens = ones* pattern, need look only at the tens in the divisor and the dividend to find the ones in the quotient. Thus, both divisor and dividend are approximated to the nearest ten. The two procedures complement each other.

When estimating a quotient, pupils need to be aware that estimates may be either too large or too small. They need to know what to do in either case—that is, try a smaller or a larger number.

$$\begin{array}{r} 7 \\ 32\overline{)218} \\ 224 \end{array}$$ 32 × 7

The quotient is too large because 224 is greater than 218. So try 6 as the quotient.

$$\begin{array}{r} 6 \ r \ 26 \\ 32\overline{)218} \\ -192 \\ \hline 26 \end{array}$$ 32 × 6

Exercises like the following can be given to help pupils identify these situations and rectify them.

The given quotient is too large.

$$\begin{array}{r} 8 \\ 32\overline{)241} \\ 256 \end{array}$$

Work to find the correct quotient and remainder.

$$\begin{array}{r} r \\ 32\overline{)241} \end{array}$$

Division with two-digit divisors is difficult because of its many component parts. In the algorithm these parts are put together. The steps in the algorithm for division examples that have quotients in the tens are outlined below.

1. Estimate the quotient: it is between 10 and 100.   43)2716

$$\begin{array}{r} 1 \times 43 = 43 \\ 10 \times 43 = 430 \\ 100 \times 43 = 4300 \end{array}$$

2. Find the tens in the quotient. Write 6 in the tens place.

$$\begin{array}{r} 6 \\ 43 \overline{)\,2716} \\ -\,2580 \\ \hline 136 \end{array}$$ +3 × 60

$4\overline{)27}$ is about 6; so 27 hundreds ÷ 4 tens is about 6 tens.

3. Find the ones in the quotient. Write 3 in the ones place.

$$\begin{array}{r} 63 \\ 43 \overline{)\,2716} \\ -\,2580 \\ \hline 136 \\ -\,129 \\ \hline 7 \end{array}$$ 43 × 3

13 ÷ 4 is about 3; 13 tens ÷ 4 tens is about 3.

4. Write the remainder with the quotient.

5. Check: Does 2716 = (63 × 43) + 7?

Concluding Comment

Success with the multiplication and division algorithms is the result of the cumulative learning from each of the prior stages. Emphasis needs to be placed on the understanding and mastery of each stage and its concepts before moving on to the next level.

Teaching the multiplication and division algorithms is difficult. Each poses unique problems. Careful attention must be given to base representations, place-value ideas, and a well-thought-out sequence of steps by which to arrive at the goal. Although everything about teaching computation is not known, the instructional suggestions and procedures presented here should lessen the difficulty that pupils have in learning the multiplication and division algorithms.

REFERENCES

Crawford, Douglas H. "An Investigation of Age-Grade Trends in Understanding the Field Axioms." (Doctoral dissertation, Syracuse University, 1964.) *Dissertation Abstracts* 25 (1965): 5728-29.

Flournoy, Frances (a). "A Study of Pupils' Understanding of Arithmetic in the Intermediate Grades." *School Science and Mathematics* 67 (April 1967): 325-33.

_____(b). "A Study of Pupils' Understanding of Arithmetic in the Primary Grades." *Arithmetic Teacher* 14 (October 1967): 481-85.

Hazekamp, Donald W. "The Effects of Two Initial Instructional Sequences on the Learning of the Conventional Two-Digit Multiplication Algorithm in Fourth Grade." (Doctoral dissertation, Indiana University, 1976.) *Dissertation Abstracts International* 37 (1977): 4933.

Knight, F. B. "An Analysis of Long Division." Twenty-ninth Yearbook of the National Society for the Study of Education, pp. 162–67. Chicago: The Society, 1930.

Payne, Joseph N., and Edward C. Rathmell. "Number and Numeration." In *Mathematics Learning in Early Childhood,* Thirty-seventh Yearbook of the National Council of Teachers of Mathematics, edited by Joseph N. Payne, pp. 125–60. Reston, Va.: The Council, 1975.

Scrivens, R. W. "A Comparative Study of Different Approaches to Teaching the Hindu-Arabic Numeration System to Third Graders." (Doctoral dissertation, University of Michigan, 1968.) *Dissertation Abstracts* 29 (1968): 839–40.

Smith, Robert F. "Diagnosis of Pupil Performance on Place-Value Tasks." *Arithmetic Teacher* 20 (May 1973): 403–8.

8

A Teaching Sequence from Initial Fraction Concepts through the Addition of Unlike Fractions

Lawrence W. Ellerbruch
Joseph N. Payne

THE teaching sequence discussed in this article provides a guide to instruction on (1) the initial concepts of fractions, (2) the addition of like fractions, (3) equivalent fractions, and (4) the addition of unlike fractions. The design of the sequence enables pupils to develop both understanding and skill and allows teachers to meet the needs of most classes by easily adjusting the amount of time spent on the material.

An important part of any instruction is ascertaining prerequisite knowledge—and teaching any prerequisites that are lacking. Before beginning computation with fractions, pupils must be able to represent fractions using concrete objects and diagrams, to recognize and use both oral and written words for fractions, and to recognize and use the symbols for fractions. Consequently, this teaching sequence begins with a thorough introduction to fraction concepts. Next, the addition of like fractions is developed, using the single mathematical model of the measure of rectangular regions. The same model is then used to demonstrate that equivalent fractions name the same part of a given region. Rectangular regions are used to develop the generalization for expressing equivalent fractions in higher terms and then used in the addition of unlike fractions.

The teaching sequence is based on results from several research studies done at the University of Michigan. This research, as well as other studies on learning fractions, is summarized by Payne (1976). The initial work on fraction concepts was tested in grades 1–6, and the material on addition was tested in grades 4–6.

Initial Fraction Concepts, Language, and Symbols

Many interpretations are needed before a child can understand fractions—and eventually rational numbers—completely. Some of these interpretations might include fractions as measures of parts of regions, segments, and three-dimensional containers; fractions as parts of a set of objects; fractions as ratios; fractions as quotients; and fractions as equivalent classes of ordered pairs of natural numbers. (See Kieren [1976] for a description of many such mathematical interpretations.)

Research as well as classroom experiences with fractions have encouraged us to base our introduction to fractions on a single mathematical model. (See Muangnapoe [1975] for data and analysis.) A measurement model seems most natural to children and most useful for the addition of fractions. In a measurement model, parts are compared to the whole unit, and a number is attached to represent the part, as shown in the examples in figure 8.1.

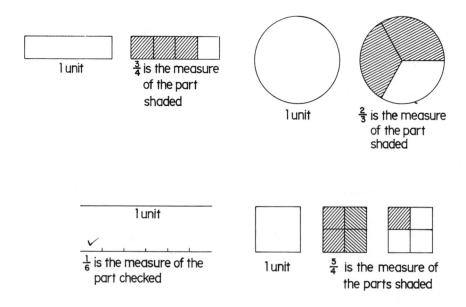

Fig. 8.1

Any of the examples in figure 8.1 requires that the picture be processed by the geometric, or spatial, portion of the brain. The number is then related to the spatial model. It is believed that this spatial representation is the essential ingredient in quantitative thinking. (See Greeno [1975] for a full discussion.)

A *rectangular* region is used here for the spatial representation because of the ease in making concrete models from strips of paper and using them. Drawing diagrams is subsequently easier to do. Furthermore, the ultimate goal of this teaching sequence—adding unlike fractions— is obtainable through the use of rectangular regions alone. (Although spatial representations such as polygonal regions, line segments, and circles are necessary for the full development of the measurement concept, these other representations are treated as extensions in this teaching sequence because they are not necessary for learning how to add fractions.)

The major steps in the sequence for developing initial fraction concepts, language, and symbols follow. The child must learn—

1. to use concrete objects and make equal-size partitions (the term *equal-size* rather than the more grammatically correct *equal-sized* is used throughout this article to reflect children's usage);
2. to recognize and use the oral names for the various-size parts;
3. to draw diagrams of the concrete objects and attach the oral names to the parts;
4. to use the concrete objects and diagrams together with the oral names to write the fraction symbols.

Each of these four major components is presented in detail.

Concrete objects and equal-size partitions

For both concrete objects and diagrams, a learner must keep in mind four essential ideas:

a) Identifying the size of the *unit*
b) Recognizing the necessity of splitting the unit into *equal-size pieces*
c) Naming the *number* of equal-size pieces *to be considered*
d) Naming the *number* of equal-size pieces *in the unit*

The following activities illustrate those that are used repeatedly throughout the process of learning about fractions:

The size of the unit. Use two candy bars or candy-bar wrappers, similar in size to the ones shown in figure 8.2. Ask, "Which candy bar would you choose? Why would you take that one?" Then compare the sizes. Discuss the importance of knowing the size of a bar (the unit size) in deciding which one to choose.

Fig. 8.2

Equal-size pieces. Hide from view two candy bars cut as in figure 8.3. "I have two candy bars, both the same size. I have cut each one into three pieces. From which candy bar would you choose a piece?" Show that the largest piece from the second candy bar is larger than any one piece from the other. Students readily recognize that a "fair cut" is made only if the pieces in each bar are the same size.

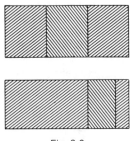

Fig. 8.3

The number of pieces. To highlight the necessity of knowing the *number* of equal-size pieces in a unit, hide from view two candy bars, each split into a different number of pieces (fig. 8.4). "This time the candy bars are the same size and each one is cut into equal-size pieces. What do you need to know to choose a bigger piece from one of the candy bars?" Pupils will readily observe that the size of a piece is determined by the number of pieces: the more pieces, the smaller each piece.

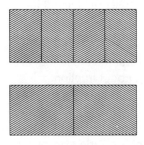

Fig. 8.4

If each equal-size unit is split into the same number of equal-size pieces, then the number of pieces chosen is the critical factor in determining which portion is more, less, or the same.

This practical introduction should take no more than ten minutes. Afterward, pupils can make "fraction strips" (fig. 8.5). Use strips of paper about four centimeters wide and twenty centimeters long. Different colors should be used to make it easy for the pupils to see "related fractions" and for the teacher to see which strips the pupils are using. Use four strips of one color for halves, fourths, and eighths; three of another color

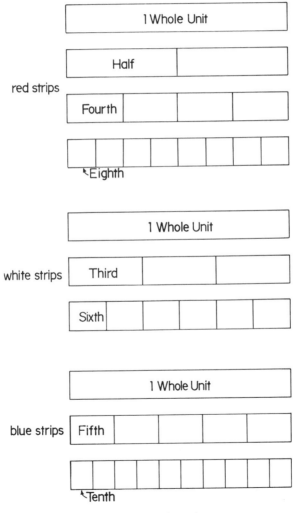

Fig. 8.5. Fraction strips

for thirds and sixths (perhaps an extra one in the upper grades to show ninths); and three of still another for fifths and tenths. All strips are to be the same size, and folds should be made parallel for ease in comparing parts. For ease in counting the pieces, a dark line should be drawn at the folds.

It is important to help pupils identify strategies for folding to obtain the desired sizes. Specific suggestions on splitting such rectangular regions into equal-size parts are found in Coxford and Ellerbruch (1975). If you do not wish to take time to teach pupils how to fold and make the fraction strips (it takes an entire class period), then duplicate the strips and have pupils cut them out. The 4 cm × 20 cm size is minimal, even with duplication.

Oral names for fraction parts

Ordinal counting is helpful at first in teaching the names for fractions. Count off objects or pupils: "First, second, third, fourth, fifth, sixth, . . ." Point out that beginning with *third* fraction names are just the same. Only the name *half* and its plural *halves* need to be learned specially.

After the names are well established and related to the appropriate strip, move ahead to naming more than one piece, having pupils indicate the amount you name. "Hold up your 'candy bar' cut into thirds. Show me one-third. Now show me two-thirds. Can you show me three-thirds?" Pose similar questions, using strips cut into fourths, fifths, and so on. Conversely, hold up strips and have the *pupils* give the oral name of the amount you indicated. Include "zero pieces" as well as fractions greater than one, such as five-fourths, so that students realize early that fractions are not always just a part of a whole thing. Pupils should note that the number of pieces is named first (one, two, three, four . . .) and the size of the pieces second (half, third, fourth, fifth . . .). Focusing attention on the order of the two numbers now will help in work on writing the fractions later.

Sufficient practice should be given to enable all pupils to respond with almost total accuracy in giving the oral names for various numbers of pieces in a unit.

Diagrams for fraction parts

Just as the strips of paper were "candy bars," the diagrams become pictures of the paper strips and, consequently, of the candy bars. By placing one of the fraction strips on the chalkboard or overhead projector and tracing around it, you can produce a picture of a unit. Draw lines in the diagram to correspond to the folds on the paper. Emphasize that you are making a picture of the object.

Pupils should then draw their own pictures of strips. Activities should be similar to those used in steps 1 and 2 above, with particular attention given to the four essential ideas related to fractions—the unit, equal-size pieces, the number of pieces considered, and the number of pieces in the unit.

Fraction symbols for concrete objects, diagrams, and oral names

With the main ideas of fractions well established and related both to concrete objects and to diagrams and with the oral names well learned, it is a simple matter to make the transition to the usual fraction symbols.

Display a strip to show three-fourths. Ask the usual four questions: "What is the unit? Are the pieces of equal size? How many pieces am I holding up (fold back one)? How many pieces in the unit?"

Then ask for the oral name when you display three-fourths: "What is the name for this amount?" Write "3 fourths" on the chalkboard. Then show that a short form of writing this is $\frac{3}{4}$. Repeat such questions for four or five other examples, following the transition from the strips of paper to the word names and finally to the fraction symbol, as shown in figure 8.6.

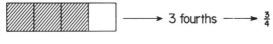

Fig. 8.6

Research has indicated that pupils make a substantial number of reversals when writing the numerator and denominator (Payne et al. 1974; Muangnapoe 1975; Choate 1975; Williams 1975; Galloway 1975; Ellerbruch 1975). When the oral names were taught before the written symbols, however, practically no reversal errors were made. This is why stress is placed on teaching the oral names before the written symbols. (Although the words *numerator* and *denominator* may be introduced as the "numberer" and "namer," a mastery of these words is not essential at this time.)

Since an understanding of fractions serves as the basis for all later operations on fractions, insist on a high level of mastery of the concepts. It is reasonable to expect children in grade 3 and above to perform at about the 90 percent level. (On an initial unit with a similar approach but with both circular and line-segment models, Muangnapoe [1975] reported that 80 percent or more of the pupils at the third-grade level attained a level of mastery of at least 80 percent.)

The introduction to fractions described above can be done in about three classroom hours with average fifth- or sixth-grade classes. For third- or fourth-grade pupils, an extra hour or two may be required. At the primary level, from seven to ten classroom hours should enable pupils to learn most of the initial concepts, language, and symbols.

Extensions

The sequence just described is designed to produce competency with fractions sufficient for work with the addition algorithm. If a mastery of fractions involving other geometric shapes is desired, circular regions and line segments are among the most important.

With circular regions, the major difficulty to be overcome is the natural tendency to attempt to cut a circle into equal-size pieces using equally spaced parallel lines. Without instruction, many children cut a circle into four parts as shown in figure 8.7 when asked to cut the circle into four equal-size parts.

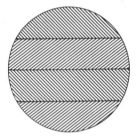

Fig. 8.7

It is no simple matter to teach them how to cut circular regions into pieces of equal size. First, begin with a circular object (coffee filters are just about the right size, readily available, and easy to fold). Fold a circular region as shown in figure 8.7. "Are the pieces the same size? How can we fold the paper over to see?" Then fold over the top and bottom pieces, making it clear that they are smaller than the middle two.

Next, tell pupils that equal-size pieces in circles are most easily made if all cuts are made on radii of the circle. This means that the center of the circle must be located. (Fold over twice, making two diameters, and the folds will intersect in the center of the circle.) Equal parts are then made by marking equal parts along the circumference of the circle. "If you want the pie cut into three equal parts, mark three equal parts on the circle. Then draw the radii." The three equal parts can be approximated by placing three objects on a circle and moving them until they appear equally spaced (fig. 8.8).

Expect to spend at least two classroom hours relating circular regions to all the key ideas of fractions. These include the connections between the circular model, the oral language, and the written symbols. A natural

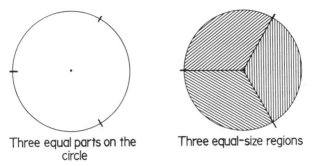

Three equal parts on the   Three equal-size regions
circle

Fig. 8.8

example for circular regions involving fractions is the clockface (see fig.
8.9).

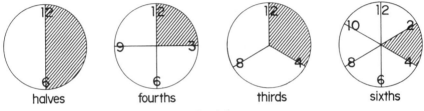

halves   fourths   thirds   sixths

Fig. 8.9

The other major extension, and the one needed to understand the use
of a ruler, is measurement along a line segment. This is best begun by
using a long straw as the basic unit because smaller parts are more visible
to pupils (fig. 8.10).

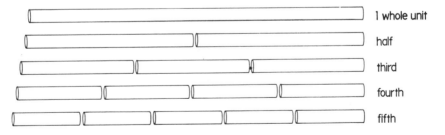

1 whole unit

half

third

fourth

fifth

Fig. 8.10

Begin with one straw taped to the chalkboard. Then cut a straw into
two equal parts and put the parts underneath, separated sufficiently for
the pupils to see each part. Do the same for thirds, fourths, fifths, and
tenths (for the centimeter ruler). Repeat the activities used with rectangular
regions, relating them to the linear units.

Initially, fractions are related to parts of a straw in a way similar to rectangular regions, with some added emphasis on the length of the straws. It is quite a bit more advanced to make the next step of attaching fractional numbers to points on a line. When fractions are related to points on a line, stress that the numbers show the amount of the straw from the beginning to the endpoint of the piece. Establishing the direction of movement is essential for the learner to see that the symbol $\frac{1}{3}$ is placed as shown in figure 8.11 and not one-third the distance from the right. The number line is difficult for children to comprehend, and care must be taken if the decision is made to include it at this time.

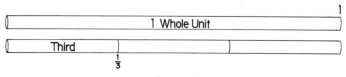

Fig. 8.11

Another extension often considered for inclusion in a fraction sequence is the set model for fractions. The research reported by Payne (1976) demonstrated clearly that there is confusion in the minds of children when the measurement model is followed by the set model. Somehow, a child will attach $\frac{1}{6}$ to the marked part of a set of six objects (see fig. 8.12) but cannot readily relate $\frac{1}{3}$ to two of the six objects. Also, when the pupils return to the measurement ideas after working with the set model, they experience either some loss of the measurement notion or some interference between the models that impairs performance with either model. Consequently, the application of fractions to sets is not included in this sequence. (For a full discussion of the problem, see Muangnapoe [1975] and Coburn [1974].)

Fig. 8.12

It is our opinion that such practical problems as finding "one-third of a dozen" may best be done through the division of whole numbers. For example, $\frac{1}{3}$ of 12 is found by dividing 12 by 3.

Addition of Like Fractions

At this point in the sequence, students should have acquired well-developed concepts of fractions, including (minimally) the application of the oral names and written symbols to measures of rectangular regions using both concrete objects and diagrams. It is then a simple matter to teach the addition of like fractions. Instruction takes only a few minutes, and with adequate practice pupils can attain mastery by the end of a single lesson. The concrete models and the oral names are used initially.

Have the pupils pair off. Instruct one member of each pair to hold up two fourths, using the fraction strips. Ask the other to hold up one fourth. Pose such questions as these:

"Are the pieces the same size?"

"How big is each piece?"

"How much is the first person holding up?"

"How much is the second person holding up?"

"How many pieces are you holding up together?"

"How much are you holding up together? Name it with a fraction."

Do a few more examples. Try having one pupil hold up a third and the other a half. Ask questions to see if they understand that since the pieces are not the same size, the resulting quantity is difficult to name.

As a transition to fraction symbols, do examples using word forms such as these:

$$
\begin{array}{cc}
2 \text{ fifths} & 3 \text{ fourths} \\
+1 \text{ fifth} & +2 \text{ fourths} \\
\hline
? \text{ fifths} & ? \text{ fourths}
\end{array}
$$

This form emphasizes the role of the size of the pieces, and it helps prevent a natural inclination to add denominators. If there is any hesitation in the students' responses, have them use their fraction strips to illustrate and solve the examples. Diagrams can be used also. Include and discuss such common errors as "2 fourths + 1 fourth = 3 eighths."

Examples with unlike denominators should be included, such as "1 fourth + 2 fifths." Pupils should mark these "can't solve." The need for equal-size pieces in both addends is thus reinforced.

When fraction symbols are finally introduced with addition, include both the concrete model and the word forms, as shown in figure 8.13. Should errors like adding denominators when adding fractions occur,

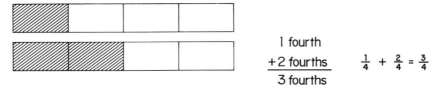

$$
\begin{array}{l}
1 \text{ fourth} \\
+2 \text{ fourths} \qquad \frac{1}{4} + \frac{2}{4} = \frac{3}{4} \\
\hline
3 \text{ fourths}
\end{array}
$$

Fig. 8.13

then state the example orally and use concrete objects to arrive at a solution.

After the pupils have attained a mastery of from 80 to 90 percent on adding like fractions, make them explicitly aware of the algorithm they are using. Ask questions that lead them to state in their own words the rules for the algorithm. If they discover the rules they are using, fine— but make sure the rules are stated. Don't assume that because students can give a correct answer they are using a correct algorithm.

The major steps to be included in the statement of an algorithm are given below. The steps are given two ways; the more informal way does not require the words *numerator* and *denominator*.

Steps for the Addition of Like Fractions

Formal Statements	*Informal Statements*
1. Are the denominators the same?	1. Are the pieces the same size? Are the bottom numbers the same?

If no, write "Can't solve."

2. Add the numerators and write the sum.	2. Add the top numbers and write the answer.
3. Draw a fraction line.	3. Draw a line underneath.
4. Write the denominator.	4. Write the bottom number.

Equivalent Fractions

We prefer next to develop equivalent fractions as a step in the general addition-of-fractions sequence. Pupils at this stage are taught to generate fractions only in higher terms. Reducing fractions to lower terms is not included at this point because of the difficulty students have with reducing and the confusion that results in the general addition case when two difficult topics are treated at the same time.

Five major steps are used in the sequence for generating equivalent fractions in higher terms. The student must—

1. recognize that *two equivalent fractions name the same amount,* or show the same measure;
2. develop and state the *generalization* that multiplying the numerator and the denominator by the same number generates an equivalent fraction;
3. find the "new" numerator, given a fraction and a "new" denominator;
4. when given a pair of fractions and a common denominator, find a "new" pair of fractions using the common denominator;

5. when given a pair of fractions, determine a common denominator and use this common denominator to find two equivalent fractions.

The first two steps are easily related to the spatial models pupils learned earlier with rectangular regions. Steps 3, 4, and 5 are perceptually more difficult, at least when initially taught. Consequently, these last three steps are taught first by using a set of rules and subsequently by relating them to the spatial models. Ellerbruch (1975) found that with this sequence pupils attained better skill with equivalent fractions and that their understanding was comparable. This is in contrast to a sequence where the spatial models are used to teach steps 3 through 5 prior to teaching rules. (In a similar study, Choate [1975] found that using concrete models simultaneously with rules for comparing fractions was not effective.)

Two equivalent fractions name the same amount

Use the fraction strips described previously to provide examples of two different fractions naming the same amount, and consequently the same number. For example, show that one-half and two-fourths cover the same amount, or same part, of a whole unit. Hence, they are *equivalent* fractions, and we can write $\frac{1}{2} = \frac{2}{4}$. The fractions $\frac{1}{2}$ and $\frac{2}{4}$ are equivalent because they both show the same amount. Many examples of related fractions are needed.

Multiplication generalization for higher terms

An effective method for building a generalization for fractions in higher terms uses a paper-folding technique (Bohan 1971). The essential idea is to relate the folding of sheets of paper to the idea of doubling, tripling, and, in general, multiplying the numerator and the denominator by the same number.

Have students take a sheet of paper and fold it to make halves. Open the sheet and shade one of the halves. Write the fraction $\frac{1}{2}$ on the chalkboard or overhead projector. Now ask pupils to fold their paper again, making sure that all parts are the same size. "Before you unfold, how many parts are now on the sheet of paper? How many of them are shaded?" Write $\frac{2}{4}$ to show the equivalent fraction. Stress the language of doubling the number of pieces in all and doubling the number considered.

Fold the paper again and repeat the questions: "How many parts now?" (Be prepared for the incorrect answer of "six.") Again, write the equivalent fraction $\frac{4}{8}$ and again stress the idea of doubling. Try folding once more, asking the same question. Begin to include questions such as, "If you double the number of pieces in the unit, what happens to the number of pieces that are shaded?" (Doubled also) (See fig. 8.14.)

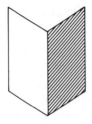  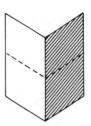

One-half is shaded.   Folding *doubles* the number   The total number of parts
                             of parts.   is *doubled;* the number of
                                                               shaded pieces is *doubled*
                                                               also.

Fig. 8.14

It may be easier to show tripling by using diagrams of rectangular regions. Shade one-half of the region; then draw two lines across the unit region to make three equal-size pieces from each original piece (fig. 8.15). Ask such questions as, "How many pieces did we start with?" (One) "What happened to the number of pieces in the unit?" (Tripled) "How many pieces are shaded now?" (Three) "What happened to the number of pieces shaded?" (Tripled)

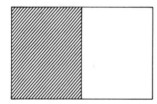 

One-half is shaded.   The total number of pieces is
                                        tripled; the number of shaded
                                        pieces is also *tripled.*

Fig. 8.15

The language of doubling and tripling leads to the generalization that equivalent fractions can be generated by multiplying *both* the numerator and the denominator by the same number. Either you or the pupils themselves should demonstrate that the equivalent fractions generated show the same portion of the rectangle.

Any application of the generalization should include examples such as these:

a) $\dfrac{2 \times 4}{3 \times 4} = $ _____   The multiplier is written in both the numerator and the denominator.

b) $\dfrac{4 \times ?}{5 \times 3} = $ _____   The multiplier is written only in the numerator or in the denominator but not in both. The multiplier must then be written in the other place and the multiplication performed to generate a new fraction.

c) $\dfrac{3 \times ?}{4 \times ?} = $ _____   The pupil is to supply the multiplier in both the numerator and the denominator to generate equivalent fractions.

Given a fraction and a "new" denominator, find the "new" numerator

In this step, pupils solve problems of the type $\frac{2}{3} = \frac{?}{12}$. They should write the multiplier for both the numerator and the denominator to show how the new numerator is determined:

$$\frac{2 \times 4}{3 \times 4} = \frac{8}{12}$$

Most pupils tend to use multiplication language to find the multiplier: "What do I multiply 3 by to get 12?" Some of them may use division language: "12 divided by 3 is 4; so multiply both by 4." Either is quite acceptable.

Given two fractions and a "new" denominator, find equivalent fractions

The pupils are given a pair of fractions with a common denominator provided. The only new step is for them to use their knowledge of equivalent fractions to generate new fractions. This step helps them become familiar with having two fractions in a single problem. They are to determine the multiplier to be used with each fraction. This multiplier is written in both the numerator and the denominator of a fraction, and the equivalent fraction is written just below.

The following format has proved successful for pupils.

Example	*Solution*
Write each fraction with a denominator of 12.	
$\dfrac{3}{4}$, $\dfrac{2}{3}$	$\dfrac{3 \times 3}{4 \times 3}$, $\dfrac{2 \times 4}{3 \times 4}$
$\dfrac{}{12}$ $\dfrac{}{12}$	$\dfrac{9}{12}$, $\dfrac{8}{12}$

Given a pair of fractions, determine a common denominator and find equivalent fractions

This step in the sequence is the most difficult. The students are given a pair of fractions, such as $\frac{3}{4}$, $\frac{5}{6}$. They must first determine a common denominator and then generate the two equivalent fractions, just as was done in the previous step. Choosing the larger of the denominators and trying successive multiples of this denominator is the algorithm suggested (from interviews with adults it was the algorithm found to be used most often [Ellerbruch 1975]).

The format and steps for presentation to pupils are illustrated as follows, using $\frac{3}{4}$ and $\frac{5}{6}$:

1. Choose the denominator that is "largest," in this case 6.

2. Write that number in the first row under the heading "Try Numbers." Multiply the number by one.

Example			*Try Numbers*
$\frac{3}{4}$	,	$\frac{5}{6}$	1. $6 \times 1 = 6$
			2.
			3.
____	,	____	4.

3. The result (6) is your first "try number." Check to see if this number can be used as a common denominator for both of the given denominators. If it can be used as a common denominator, use it; otherwise, generate the next try number (6 cannot be used, since no whole number times four equals six).

4. To generate the next try number, write the "largest" denominator on the next line (in this case the second line) and multiply it by two. This will give the next try number.

Example			*Try Numbers*
$\frac{3}{4}$	,	$\frac{5}{6}$	1. $6 \times 1 = 6$
			2. $6 \times 2 = 12$
			3.
____	,	____	4.

Check the try number as in step 3. Continue using step 4 until a common denominator is found (12 will work as a common denominator because $4 \times \underline{3} = 12$ and $6 \times \underline{2} = 12$).

5. Use the try number as a common denominator and complete the problem as was done in step 4 of the sequence where the denominator was given.

Example		*Try Numbers*
$\dfrac{3 \times 3}{4 \times 3}$,	$\dfrac{5 \times 2}{6 \times 2}$	1. $6 \times 1 = 6$ 2. $6 \times 2 = 12$
$\dfrac{9}{12}$,	$\dfrac{10}{12}$	3. 4.

The problems should not be separated into special types, such as simple multiples or relatively prime denominators or any other special cases. If certain types of fractions are treated as special cases, pupils will tend to develop special algorithms for each case. The algorithm used here will generate the lowest common denominator if the pupils do not make computation errors.

Difficulties with this major step in the sequence are most often related to the multiplication or division of whole numbers. Make sure that pupils have and use the needed thinking strategies for both operations with whole numbers.

After they demonstrate success with the algorithm, begin to remove the "crutches" in the example format. Eliminate first the numbers 1, 2, 3, and 4, and then the heading "Try Numbers." Next, the helping lines under the fractions can be omitted because they become natural in writing the equivalent fractions. If pupils have difficulty with the algorithm, remind them that they may still use the crutches if they find them helpful.

The Addition of Unlike Fractions

The only new step for the general case of adding fractions is to use a plus sign and add the equivalent fractions with a common denominator. This step will not require much practice. Should there be difficulties, return to the "try numbers" format for equivalent fractions.

After the pupils are able to use the addition-of-fractions algorithm, relate the algorithm to the rectangular region model. This can be done by manipulating materials to correspond to the steps in the algorithm; that is, establish a connection between the steps in the paper-and-pencil algorithm and the manipulation of the concrete materials. One way to illustrate $\frac{3}{4} + \frac{5}{6}$ is shown in figure 8.16.

Relating the spatial models and the rules will reinforce the idea that the

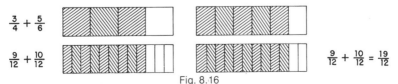

Fig. 8.16

addition of fractions can be processed mentally either way. There are times when the spatial model is preferred (for mental work and simpler problems) and times when the rule is needed (for complicated examples, particularly).

Time Lines

The following suggestions for the necessary amount of time to teach the different portions of the sequence are based on experience in teaching classes of average to above-average ability. For slower classes, more time will very likely be needed. The effective pacing of the content depends on the fine skill of a sensitive teacher. (See Williams [1975] for results on using the sequence on initial concepts in remedial situations.)

Primary grades

Primary school children can complete the work on initial concepts and language in seven to ten days. The instruction during this time will emphasize the concrete materials, the oral language, and possibly the diagrammatic representation. Galloway (1975) provides more information on the implementation of the initial work in the primary grades.

Fourth grade

The initial work on fraction concepts should be allocated approximately five days, but more time should be allowed if the pupils are not performing at a mastery level of 90 percent. The addition of like fractions can be accomplished in one day. The portion of the sequence devoted to equivalent fractions may require up to five days. In fact, there is a good chance that many fourth graders will not be able to complete this part of the sequence successfully. If the students *can* complete the entire equivalent fraction sequence, they will be able to do the addition of unlike fractions and complete that part in one or two days.

Fifth through eighth grades

The initial work on fraction concepts can be done in approximately three days and the addition of like fractions in one day. Equivalent fractions will take three to five days. The addition of unlike fractions can be accomplished in one day.

The expectation is that the mathematics period is from forty-five minutes to one hour each day. Adjustment will have to be made in the number of days allocated if less time is used each day.

Summary

This teaching sequence is flexible enough to be broken up into smaller or larger segments to meet the individual needs of students and a variety of classroom settings. It began with instilling a thorough understanding of fractions using concrete models and rectangular regions. The subsequent algorithms for the addition of like fractions, equivalent fractions, and the addition of unlike fractions were done with the assumption clearly in mind that students do understand the initial work with fractions. The sets of experiences given here are minimal for providing a maximum chance for students to acquire both understanding and skill with the addition of fractions.

REFERENCES

Bohan, Henry J. "A Study of the Effectiveness of Three Learning Sequences for Equivalent Fractions." (Doctoral dissertation, University of Michiagn, 1970.) *Dissertation Abstracts International* 31 (1971): 6270A. (University Microfilms no. 71-15, 100)

Choate, Stuart A. "The Effect of Algorithmic and Conceptual Development for the Comparison of Fractions." (Doctoral dissertation, University of Michigan, 1975.) *Dissertation Abstracts International* 36 (1975): 1410A. (University Microfilms no. 75-20.316)

Coburn, Terrence G. "The Effect of a Ratio Approach and a Region Approach on Equivalent Fractions and Addition/Subtraction for Pupils in Grade Four." (Doctoral dissertation, University of Michigan, 1973.) *Dissertation Abstracts International* 34 (1974): 4688A–89A. (University Microfilms no. 74-3559)

Coxford, Arthur F., and Lawrence W. Ellerbruch. "Fractional Numbers." In *Mathematics Learning in Early Childhood,* Thirty-seventh Yearbook of the National Council of Teachers of Mathematics, pp. 191–203. Reston, Va.: The Council, 1975.

Ellerbruch, Lawrence W. "The Effects of the Placement of Rules and Concrete Models in Learning Addition and Subtraction of Fractions in Grade Four." Unpublished doctoral dissertation, University of Michigan, 1975. (University Microfilms no. 76-9389)

Galloway, Patricia W. "Achievement and Attitude of Pupils toward Initial Fractional Number Concepts at Various Ages from Six through Ten Years and of Decimals at Ages Eight, Nine and Ten." Unpublished doctoral dissertation, University of Michigan, 1975.

Greeno, James G. "Cognitive Objectives for Instruction: Knowledge for Solving Problems and Answering Questions." In *Cognition and Instruction,* edited by D. Klahr. Hillsdale, N.J.: Lawrence Erlbaum Associates, 1975.

Kieren, Thomas E. "On the Mathematical, Cognitive, and Instructional Foundations of Rational Numbers." In *Number and Measurement, Papers from a Research Workshop,* edited by Richard A. Lesh, pp. 101–44. Columbus, Ohio: ERIC/SMEAC, 1976.

Muangnapoe, Chatri. "An Investigation of the Learning of the Initial Concept and Oral/Written Symbols for Fractional Numbers in Grades Three and Four." (Doctoral dissertation, University of Michigan, 1975.) *Dissertation Abstracts International* 36 (1975): 1353A–54A. (University Microfilms no. 75-20.415)

Payne, Joseph N. "Review of Research on Fractions." In *Number and Measurement, Papers from a Research Workshop,* edited by Richard A. Lesh, pp. 145–87. Columbus, Ohio: ERIC/SMEAC, 1976.

Payne, Joseph N., James G. Greeno, Stuart A. Choate, Chatri Muangnapoe, Hazel B. Williams, and Lawrence W. Ellerbruch. "Initial Fraction Sequence." Mimeographed. Ann Arbor, Mich.: University of Michigan, 1974.

Williams, Hazel B. "A Sequential Introduction of Initial Fraction Concepts in Grades Two and Four and Remediation in Grade Six." Ed.S. research report, School of Education, Mimeographed. Ann Arbor, Mich.: University of Michigan, 1975.

9

Assessing the Development of Computation Skills

George W. Bright

DID you know that your federal, state, and local tax dollars have paid for a variety of assessment data on student achievement? Do you know what has been learned about achievement? In this article student performance on some aspects of computation is traced, and as a first step in the process of diagnosis and remediation, some areas of needed improvement in skill development are identified.

Although the total amount of assessment data is quite large, the amount of data related to some specific skills is too small to permit adequate analysis of the development of those skills. Thus not every aspect of computation can be considered in this article. Many recent assessment projects, perhaps as a response to calls for moving "back to basics," have focused on computation with whole numbers, fractions, and decimals. This article reflects a similar emphasis.

It is important to examine the data in order to find trends that indicate how well students can compute. Patterns in the development of skills must be identified before effective changes can be suggested for the ways in which those skills are introduced, maintained, and—as a last resort—remediated. Relating the various kinds of assessment data will reveal some patterns that no single set of data could be expected to substantiate.

The data that will be used come from three sources: the National Longitudinal Study of Mathematical Abilities (NLSMA), the National Assessment of Educational Progress (NAEP), and recent state and local assessment programs. These sources were selected primarily because the data are provided on an item-by-item basis rather than simply as average test scores or grade-equivalent scores for entire samples or populations. Knowing the percentages of students who correctly respond to specific exercises related to specific skills is needed for identifying skills that are especially easy or especially difficult.

The main problem in attempting to synthesize the wide variety of assessment data is that each group that has assessed the development of skills has had unique goals, used a unique population, and tested by a unique procedure. In Georgia, for example, all public school students in grades 4, 8, and 11 participated in the statewide assessment program (Georgia 1976). In contrast, NAEP used sophisticated sampling techniques to gather representative data on nine-, thirteen-, and seventeen-year-olds, as well as adults aged twenty-six through thirty-five, for the first national mathematics assessment (Moore, Chromy, and Rogers 1974). NLSMA, a much earlier effort, used available rather than representative schools; as a result, about 70 percent of that sample came from western and northern Atlantic states (Carry and Weaver 1969; McLeod and Kilpatrick 1969).

In order to compare the various data sets accurately, one should also note that the NAEP data are reported by age rather than by grade. For example, although its sample of nine-year-olds included mostly fourth graders, almost 25 percent were in the third grade (NAEP 1975, p. xiii). Thus, whereas it is more sensible to compare the NAEP data for nine-year-olds with other data for fourth grades rather than for third grades, the NAEP data can be expected to show a less complete development of skills than those fourth-grade data (Carpenter et al. 1975 [a]). Similar caution must be exercised for other age levels of the NAEP data.

Despite differences in the ways the data were collected, analyzed, and reported, comparisons of the results within the broad perspective of an examination of the development of computation skills are needed. Only by making such comparisons can it be determined whether recurring patterns in the data exist. If so, then because of the diversity of sources, one can have faith that those patterns truly reflect the actual state of development.

Some questions that will be addressed in this article follow. Do difficult exercises have common characteristics? Are some skills usually better developed than others? Are skills that are introduced in primary grades better developed than skills introduced in the intermediate grades? Are the earlier-taught skills in fact retained? If patterns in skill development can be identified, then supplementary instructional strategies can be designed for the more difficult areas.

Comparisons Based on
Grade-Equivalent and Total Test Scores

Most of the published data on the development of mathematical skills and competencies are in the form of either grade-equivalent or total test scores. Although these data are of limited use in identifying particular weaknesses in the development of skills, they do point to general trends in performance. The studies that are cited below seem to be representative of the available literature.

Beckmann (1969; 1970) compared mathematical competencies of large samples of eighth- and ninth-grade Nebraska students in 1951 and 1966. His test of 109 items assessed many competencies besides computation —for example, concept attainment, geometry, and problem solving. The comparisons favored the 1966 students. At the eighth-grade level the 1951 and 1966 scores were 45.7 and 54.9, respectively; at the ninth-grade level the scores were 54.3 and 61.1, respectively.

Roderick (1973) used data from the Iowa Test of Basic Skills to compare performance in 1973 with that in 1936 and 1951. For both whole-number and fractional-number computation there were significant differences ($p < .01$) at grades 6 and 8 favoring the 1936 data over the 1973 data (there were no significant differences between the 1951 and 1973 data). A close examination of the problems used in 1936 and repeated in 1973, however, suggests that part of this result is attributable to differing instructional practices. The 1936 problems are typically multi-digit (more than four-digit) problems, which may have been representative of the 1936 curriculum but which are certainly atypical of the 1973 curriculum.

Hungerman (1977), in a comparison of computation skills from 1965 to 1975, related the performance of sixth graders from a southeastern Michigan city *(N = 305 in 1965, N = 386 in 1975)* on the California Arithmetic Test: Fundamentals. The analyses of the four subtests (twenty items each) revealed some very interesting patterns of performance. The 1965 scores were significantly higher on the addition subtest as a whole; but the 1975 performance was higher on seven of seven whole-number items, whereas the 1965 performance was higher on six of seven fraction items and three of three decimal items. The 1965 scores were significantly higher on the subtraction subtest as a whole; but the 1975 performance was higher on seven of seven whole-number items, whereas the 1965 performance was higher on four of seven fraction items and three of three decimal items. There were no significant differences in the scores on the multiplication subtest; but the 1975 performance was higher on nine of nine whole-number items, whereas the 1965 performance was higher on six of eight fraction items. The 1975 scores were significantly higher on the division subtest as a whole, with the 1975 performance higher on ten of ten whole-number items, four of eight fraction items, and two of two decimal items. The clear advantage of the 1975 students with whole-number computation and of the 1965 students with fractional-number computation seems to demand further study. (Hungerman will report additional analyses at a later time.)

Niemann (1973), in a status study of seventh-, eighth-, and ninth-grade students, reported that the eighth-grade students significantly outscored the seventh-grade students, but there was no appreciable difference between the scores of eighth- and ninth-grade students. Again, many competencies were measured, but one computation competency, labeled "arithmetic of positive rational numbers," yielded significant differences for ninth graders over eighth graders ($p < .05$) and over seventh graders ($p < .001$). The particular items composing this subscale were not displayed; so the skill may have been contaminated by translation or problem-solving skills.

The Missouri state assessment (1971) reported the results of a sample of 10 percent of the fourth and sixth graders. The participants were taken from available schools (i.e., volunteers). For the computation subscale, the average grade-equivalent scores were 4.9 and 6.7. South Carolina (Finch 1975) tested all fourth and seventh graders and 11 percent of the ninth and eleventh graders. The mean grade-equivalent scores for computation were 3.1, 5.9, 7.9, and 9.7, respectively.

It seems important to note that the data do suggest improvement in performance across time and particularly across grade levels, presumably because there has been more instruction and more time for practice. This observation is important because it supports the conclusion that students do learn. It is impossible, however, to identify *specific* areas in which improvement has occurred; the global picture may, in fact, mask decreasing performance in some areas. State and national assessment data are just now beginning to provide the details that have previously been hidden in results reported as grade-equivalent or total test scores.

Organization of Data

The entries in each table are the percentages of students who correctly answered the given items. The following symbols have been used to shorten references to the data sources:

Symbol	Data Source
CT	Murphy (1968). (All percentages are approximate, since they were read from a graph; students were from school districts in Connecticut [CT].) Sample items used with permission of the author.

MI   Michigan State Assessment (Zoet 1974; Beardsley et al. 1975; Coburn, Beardsley, and Payne 1975). Sample items reprinted by permission of the Michigan Department of Education and Educational Testing Service, copyright owners.

NAEP   National Assessment of Educational Progress (1975)

NC   North Carolina State Assessment (1974)

NLSMA   National Longitudinal Study of Mathematical Abilities (1968 [a]; [b]; [c]; [d]). Used with permission of the School Mathematics Study Group.

Each table was compiled to summarize the available information on student performance with respect to one area of computation skill. The data are synthesized by a search for two kinds of patterns. First, in each row of the table do the percentages increase, decrease, stabilize, or show no consistent pattern? That is, as students get older, do they perform better, worse, just as well, or is there no pattern at all? Such patterns help to reveal the progress that students make as they move through school. Stabilization at a high level, for example, would suggest a mastery of the skill by most students. Second, what are the patterns in the data in each column? The data were collected at different times and with different groups, exercises, and testing procedures. Nonetheless, since the items are arranged roughly in order of increasing complexity, the levels of performance probably will decrease in each column.

Addition and Subtraction

The first area to be discussed is the addition and subtraction of whole numbers without regrouping (table 9.1).

These data (table 9.1) support the conclusion that performance begins at a high level and stabilizes above the 90 percent level before junior high school. That is, more than 90 percent of the students can correctly add or subtract two numbers when regrouping is not required. The level of performance decreases somewhat when larger numbers are involved, but that is to be expected, since such problems allow more chances for mistakes. Similar levels of achievement are shown across all the data sources. This suggests that performance is fairly stable and is independent of geography, year of testing, and testing procedures.

The addition (table 9.2) and subtraction (table 9.3) of whole numbers involving regrouping is more complex, since an understanding of place value is critical. The same two kinds of patterns in the data will be important, however.

There seems to be more fluctuation in the data of tables 9.2 and 9.3 than in the data of table 9.1, especially at the lower grades. As one might

TABLE 9.1
ADDITION AND SUBTRACTION, NO REGROUPING

Item	Source	Grade										
		2	3	4	5	6	7	8	9	10	11	12
32 24 + 40	CT	60	93	90	93							
43 + 6	CT					92	87	92	91	92	97	98
three-digit addition	NC		83[a]									
29 − 18	CT	60	90	98	98							
96 − 64	CT					81	91	91	91	95	94	92
973 − 201	NLSMA				85		93					
three-digit subtraction	NC		85[a]									
489,263 − 265,051	MI				67							

[a]Average performance for two items.

TABLE 9.2
ADDITION WITH REGROUPING

Item	Source	Grade (Age)[a]										
		2	3	4(9)	5	6	7	8(13)	9	10	11(17)	12 Adult
38 + 19	NAEP			79				94			97	97
75 + 8	CT	22	76	82	82							
103 + 7	NLSMA			82			83					
96 + 85	NLSMA			84			93					
434[b] + 268	MI						94[c]					
three-digit addition	NC		66[d]									
378 63 + 504	NLSMA				76		95					
452 137 + 245	CT	22	72	81	85							
229 5084 63 + 1381	CT					77	81	90	79	88	80	86

TABLE 9.2—*Continued*

Item	Source	Grade (*Age*)[a]										
		2	3	4(9)	5	6	7	8(13)	9	10	11(17)	12 *Adult*
2,371[b] 50,452 + ___938	MI							88(87)[c,e]				

[a]NAEP data are reported by age rather than grade.

[b]This item is one of five similar items used to measure the same objective.

[c]The percentage given is the percentage of students who correctly answered at least four of five similar items.

[d]Average performance for two items.

[e]The data are from 1973 (1974).

TABLE 9.3
SUBTRACTION WITH REGROUPING

Item	Source	Grade (*Age*)[a]									
		4(9)	5	6	7	8(13)	9	10	11(17)	12	*Adult*
36 − 19	NAEP	55				89			92		92
73 − 66	NLSMA	83			93						
455 − 166	NLSMA	65			79						
726 − 349	CT			68	88	91	83	84	88	92	
506 − 223	MI	51									
800[b] − 277	MI				82[c]						
1054 − 865	NAEP	27				80			89		90
5060 − 2548	NLSMA	40			74						

[a, b, c]See notes on table 9.2.

hope, the scores increase across grades, with a couple of minor exceptions. There is also some decrease in performance as problems become more complex, though the differences in tables 9.2 and 9.3 are slight. By about grade 8, performance has stabilized at about 90 percent.

In general the data suggest that improvement in computation skill in addition and subtraction continues after the point at which most teachers probably expect mastery (Carpenter et al. 1975 [b]). Although the CT data, grades 9-12, do not fit this pattern completely, it is unclear what caused the fluctuations in these data. The data also suggest, as could be expected, that problems requiring regrouping are mastered later and not as well as problems involving no regrouping. Further, it seems that multiple regrouping in subtraction is more difficult than in addition. It seems necessary, therefore, that a careful look be taken at the ways subtraction with regrouping is introduced, practiced, and retaught to students who have trouble attaining mastery.

Multiplication and Division

Data dealing with performance on multiplication items are considerably more sketchy than similar data on addition and subtraction items. Since multiplication is introduced later than addition and is a more complicated process, one would expect the performance levels to be somewhat lower. However, a search will be made for the same kinds of patterns in the data—increased performance across grades and decreased performance across complexity of problem. Of special interest will be noticeable drops in performance for specifically classifiable kinds of problems—for example, problems involving zeros.

Several observations can be made about the data in table 9.4. First, the level of performance is generally lower than for addition and subtraction. Second, data for grades 4 and 6 seem to fluctuate so much that it would be pointless to attempt any detailed interpretation. Third, performance

TABLE 9.4
MULTIPLICATION

Item	Source	Grade (Age)[a]									
		4(9)	5	6	7	8(13)	9	10	11(17)	12	*Adult*
3×200[b]	MI					88(91)[c,d]					
56 $\times$ 3	NLSMA	40			97						
38 $\times$ 9	NAEP	25				83			88		81
67×7[b]	MI					82(87)[c,d]					
36 $\times$ 12	NLSMA	6			92						
3×604	MI	55									
809 $\times$ 47	CT			46	67	72	62	74	75	74	

a, b, cSee notes on table 9.2.
dThe data are from 1973 (1974).

seems to stabilize at grade 7. Fourth, test scores do not seem to improve significantly after grade 7.

The data on division items are more extensive (table 9.5). For seventh-grade data on division problems with single-digit divisors, there is a pattern of decreasing performance as the number of digits in the dividend increases. Too, there is a marked increase in performance between grades 4 and 7. This corresponds with the typical emphasis on multiplication and division found in fourth- and fifth-grade textbooks. Multiple-digit divisors cause more problems for students, though the available data do not allow a determination of whether mastery is attained in high school.

TABLE 9.5
DIVISION

Item	Source	Grade (Age)[a]									
		4(9)	5	6	7	8(13)	9	10	11(17)	12	Adult
3$\overline{)69}$	NLSMA	80			94						
7$\overline{)91}$	NLSMA	61			88						
7$\overline{)91b}$	MI					71(76)[c,d]					
5$\overline{)125}$	NAEP	15				89			93		93
6$\overline{)846}$	NLSMA	53			84						
6$\overline{)7356}$	NLSMA	32			78						
9$\overline{)4527}$	CT			42	72	83	83	82	81	92	
5 digits ÷ 1 digit, no remainder	NAEP	5				67			78		77
3 digits ÷ 2 digits, no remainder	NAEP					66			85		
32$\overline{)9792}$	NLSMA		44		67	69					
24$\overline{)482}$	NLSMA		61		66	72					
500$\overline{)3029}$	NLSMA		60		84	90					

[a,b,c]See notes on table 9.2.
[d]The data are from 1973 (1974).

Fractions and Decimals

In interpreting data on the addition and subtraction of fractions, one should keep in mind the three relationships that can exist among the denominators: (1) they are identical; (2) one is a multiple of all others; (3)

the lowest common denominator is greater than each one. The list is in order of increasing complexity and hence in order of increasing probable difficulty for students. Since the multiplication of fractions does not depend on finding a common denominator, performance levels may be higher than for addition when the lowest common denominator does not appear in the problem. There were not enough data available on the division of fractions to allow an analysis. (See table 9.6.)

TABLE 9.6
FRACTIONS

Item	Source	Grade (Age)[a]											
		1	2	3	4	5	6	7	8(13)	9	10	11(17)	12
$\frac{1}{4} + \frac{2}{4}$	NLSMA				17			89					
$\frac{2}{3} + \frac{4}{3}$[b]	MI							65(66)[c,d]					
$2\frac{1}{2} + 1\frac{1}{2}$	CT						51	78	85	91	89	92	96
$4\frac{1}{3} + 2\frac{1}{3}$[b]	MI							61(56)[c,d]					
$\frac{7}{9} - \frac{5}{9}$	CT	0	1	4	21	52							
$\frac{14}{16} - \frac{5}{16}$	NLSMA							90	92			86	
$3\frac{5}{6} - \frac{1}{6}$[b]	MI							63(67)[c,d]					
$\frac{3}{8}$ $+\frac{1}{2}$	NLSMA				62				81				
$6\frac{1}{4}$ $1\frac{5}{8}$ $+4\frac{1}{2}$	CT						33	69	77	77	77	83	81
$\frac{11}{12} - \frac{5}{6}$	NLSMA							82	87			84	
$\frac{7}{8} - \frac{1}{2}$	MI							33					
$\frac{9}{10} - \frac{101}{1000}$	NLSMA							45	55			56	

TABLE 9.6—*Continued*

Item	Source	1	2	3	4	5	6	7	8(13)	9	10	11(17)	12
$\frac{1}{2} + \frac{1}{3}$	NAEP								42			66	
$\frac{1}{2} + \frac{1}{3} + \frac{1}{4}$	MI							46					
$10\frac{1}{4}$ $-7\frac{2}{3}$	CT						18	53	67	53	53	51	54
$\frac{1}{2} \times \frac{1}{4}$	NAEP								62			74	
$\frac{3}{4} \times \frac{8}{9}$	NLSMA							67	74				
$5 \times \frac{2}{25}$	NLSMA							65	76				
$\frac{1}{6}$ of 30	CT						62	79	90	91	89	95	98
$\frac{7}{9} \times 4$	NLSMA							61	70				
$6 \times 3\frac{7}{8}$	CT						2	21	36	34	33	32	31

a, b, cSee notes on table 9.2.
dThe data are from 1973 (1974).

Performance seems to stabilize at about grade 8 (age thirteen), and there is noticeable improvement in performance from grade 6 to grade 8. As might be expected, when the lowest common denominator does not appear in the problem, students have considerably more trouble than when it does appear. Their skills with multiplying fractions seem to lie somewhere between these two levels. It is impossible to determine from the data, however, whether their difficulties result from a lack of understanding of the processes, an inability or inattention to "reducing" the answers to lowest terms, or a lack of practice. Remediation should include, first, mastering the writing of equivalent fractions; second, justifying the need for writing fractions with like denominators before adding or subtracting; and third, developing models for the multiplication of fractions.

It should be noted in passing that the multiplication algorithm, because it can be performed without writing the fractions with a common denominator, is mechanically an easy one. Conceptually, however, multiplying fractions is more complex than adding them. As noted above, this hierarchy of difficulty is supported empirically. Apparently the mechanical difficulties of adding fractions without a common denominator outweigh the conceptual difficulties of multiplication.

Computation with decimal fractions is related more to work with whole numbers than to similar work with common fractions. Typically, however, decimal fractions are not introduced until after the introduction of common fractions; so the delay and decreased exposure may work against the mastery of skills. (See table 9.7.)

TABLE 9.7
DECIMALS

Item	Source	Grade (Age)[a]											
		1	2	3	4(9)	5	6	7	8(13)	9	10	11(17))	12 Adult
$62.00 + 5.30	CT	1	2	10	45	65							
3.09 10.00 9.14 + 5.10	NAEP				40				84			92	86
0.8 + 0.5	NLSMA					53			73				
62.3 − 4.9	NLSMA						69		75			78	
If 23.8 is subtracted from 62.1, the result is	NAEP								61			78	74
4.95 × 3	CT					62	85	87	87	95	86	90	
93.6 ÷ 3	NLSMA						88	92		94			
6.23 × 12.7	CT					20	48	62	59	59	64	71	
2.9⟌308.85	CT					0	18	47	51	47	52	53	

[a]NAEP data are reported by age rather than grade.

Performance seems to stabilize by about grade 8, though the level of performance may not be as high as some would like. Except for the very complex items, the level is somewhat higher than the levels of performance shown in table 9.6. Again performance improves between grades 6 and 8. The multiplication and division of two decimal fractions suffers the same fate as the addition of common fractions for which the lowest common denominator does not appear in the item—such problems are difficult. In light of the data presented previously for the multiplication of whole numbers, at least part of the difficulty would appear to be attributable to the placement of the decimal point in the answer. Perhaps instruction in this area would be more effective if it were accompanied by a greater emphasis on estimation.

Synthesis

Overall, several patterns in the data seem to support clear conclusions. First, there is general improvement in performance across grades. This result is not unexpected, and it is consistent with the results of the grade-equivalent studies discussed earlier. Second, the levels of performance decrease as the items become more complex. Third, performance tends to stabilize. For the areas discussed in this article, stabilization seems to occur during the junior high school years. Two questions come to mind: Could stabilization be reached earlier, and if so, should resources be expended to accomplish this? Fourth, stabilization of performance for whole-number computation occurs earlier and at a higher level than for fractional-number computation. Perhaps this is a practice effect that reflects the introduction of whole-number computation before fractional-number computation. Fifth, for all computation skills considered, there is no decline—or at least no important decline—in the performance of adults in comparison to that of high school students. In the context of improvement of skill performance across grades, this suggests that once skills are mastered, they are not forgotten.

One contribution of the analysis in this article over previous discussions of the attainment of computation skills is that several specific areas of difficulty have been identified; this makes possible the planning of remediation efforts in these areas. Of perhaps greater importance is the observation that computation skills are not acquired on the basis of initial instruction. Instruction over several years is needed to reach stability, and in every area examined there is still room for improvement in students' acquisition of computation skills. Greater attention, then, needs to be given to the design of initial instruction to make it more effective.

It is important to note that the data presented here refute the notion that students generally do not acquire basic computation skills. In fact, some skills (e.g., addition and subtraction without regrouping) are almost universally acquired, whereas others (e.g., division of decimal fractions) are not. Any meaningful discussion of the performance of students in basic computation skills must be a discussion of specific skills rather than skills in general.

An adequate evaluation of the evolution of achievement in computation skills requires two things: (1) a context for discussion and (2) data from many years of testing. Progress for reaching the best context for discussion has been made by Trafton and Suydam (1975), who present ten tenets on the teaching of computation. These tenets deserve serious consideration, but more contemplation of this context is needed, especially in light of the advent of inexpensive calculators. This article is a step in providing usefully organized data on the acquisition of computation skills. The data are too recent, however, to allow enough perspective for complete interpretation. It is through the collection, analysis, and interpretation of data from many years that those skills that need remediation can be identified. Within such a context, teachers must be alert for specific needs of their own pupils. Longitudinal data, however—whether for large populations or individual pupils—are an effective means of determining what skills need remediation as well as whether remediation efforts are successful. The challenge is to alter all phases of instruction—initial teaching, practice, and reteaching—from the base of current data to help students master skills efficiently.

REFERENCES

Beardsley, Leah M., Terrence G. Coburn, Alan A. Edwards, and Joseph N. Payne. *Michigan Educational Assessment Program Mathematics Interpretive Report, 1974, Grade 4 and 7 Tests.* Birmingham, Mich.: Michigan Council of Teachers of Mathematics, 1975.

Beckmann, Milton William. "Ninth Grade Mathematical Competence—15 Years Ago and Now." *School Science and Mathematics* 69 (1969): 315–19.

_____. "Eighth Grade Mathematical Competence—15 Years Ago and Now." *Arithmetic Teacher* 17 (1970): 334–35.

Carpenter, Thomas P., Terrence G. Coburn, Robert E. Reys, and James W. Wilson (a). "Results and Implications of the NAEP Mathematics Assessment: Elementary School." *Arithmetic Teacher* 22 (1975): 438–50.

_____(b). "Subtraction: What Do Students Know?" *Arithmetic Teacher* 22 (1975): 653–57.

Carry, L. Ray, and J. Fred Weaver. *Patterns of Mathematics Achievement in Grades 4, 5, and 6: X-Population.* Palo Alto, Calif.: School Mathematics Study Group, 1969. (ERIC no. ED 044 283)

Coburn, Terrence G., Leah M. Beardsley, and Joseph N. Payne. *Michigan Educational Assessment Program Mathematics Interpretive Report, 1973, Grade 4 and 7 Tests.* Birmingham, Mich.: Michigan Council of Teachers of Mathematics, 1975.

Finch, John H. "Fall 1974 South Carolina State Testing Program." Columbia, S.C.: South Carolina State Department of Education, 1975. (ERIC no. ED 110 476)

Georgia Department of Education. "Criterion-referenced Tests in Georgia Schools, Some Questions and Answers." Atlanta: The Department, 1976.

Hungerman, Ann D. "1965–1975: Achievement and Analysis of Computation Skills Ten Years Later." Paper read at the 55th Annual Meeting of the National Council of Teachers of Mathematics, April 1977. Multilithed.

McLeod, Gordon K., and Jeremy Kilpatrick. *Patterns of Mathematics Achievement in Grades 7 and 8: Y-Population.* Palo Alto, Calif.: School Mathematics Study Group, 1969. (ERIC no. ED 084 114)

Missouri State Department of Education. "A Summary of Fourth and Sixth Grade Basic Skills." Jefferson City, Mo.: The Department, 1971. (ERIC no. ED 077 990)

Moore, R. Paul, James R. Chromy, and W. Todd Rogers. *The National Assessment Approach to Sampling.* Denver, Colo.: National Assessment of Educational Progress, 1974. (ERIC no. ED 099 416)

Murphy, George M. "Arithmetics Anonymous (An Analysis of Computational Mathematical Achievement in Six Towns)." Winsted, Conn.: Cooperative Educational Services Center, 1968. (ERIC no. ED 033 769)

National Assessment of Educational Progress (NAEP). *Math Fundamentals, Selected Results from the First National Assessment of Mathematics.* Denver, Colo.: NAEP, 1975.

National Longitudinal Study of Mathematical Abilities (a). *Description and Statistical Properties of X-Population Scales.* Palo Alto, Calif.: School Mathematics Study Group, 1968. (ERIC no. ED 044 280)

———(b). *Description and Statistical Properties of Y-Population Scales.* Palo Alto, Calif.: School Mathematics Study Group, 1968. (ERIC no. ED 044 310)

———(c). *X-Population Test Batteries, Parts A and B.* Palo Alto, Calif.: School Mathematics Study Group, 1968. (ERIC no. ED 044 277)

———(d). *Y-Population Test Batteries, Parts A and B.* Palo Alto, Calif.: School Mathematics Study Group, 1968. (ERIC no. ED 044 278)

Niemann, Donald F. "A Study of the Degree to Which 7th, 8th, and 9th Grade Students Have Obtained Minimum Mathematical Competencies and Skills as Recommended by the National Council of Teachers of Mathematics." (Doctoral dissertation, University of Nebraska, 1973.) *Dissertation Abstracts* 34 (1973): 7522A. (University Microfilms no. 74-13 006)

North Carolina State Department of Public Instruction. "Mathematics Grade 3 State Assessment of Educational Progress in North Carolina, 1973–74." Raleigh, N.C.: The Department, 1974. (ERIC no. ED 108 974)

Roderick, Stephen A. "A Comparative Study of Mathematics Achievement by Sixth Graders and Eighth Graders, 1936 to 1973, 1951–55 to 1973, and 1965 to 1973." (Doctoral dissertation, University of Iowa, 1973.) *Dissertation Abstracts* 34 (1973): 5601A–5602A. (University Microfilms no. 74-7423)

Trafton, Paul R., and Marilyn N. Suydam. "Computational Skills: A Point of View." *Arithmetic Teacher* 22 (1975): 528–37. (See p. ix of this yearbook.)

Zoet, Charles J. *Michigan Fourth and Seventh Grade Student Performance in Mathematics, 1970–1973.* Birmingham, Mich.: Michigan Council of Teachers of Mathematics, 1974.

10

Diagnosing Computational Difficulty in the Classroom

James E. Inskeep, Jr.

CHILDREN'S difficulties with computation can be classified according to (1) those dealing with the basic facts and (2) those concerning the algorithm. Some children need help with both, but giving help involves careful diagnosis. The areas of difficulty and the specific processes that lead to error need to be uncovered. With good diagnosis, appropriate remedial measures can be taken. This article presents some procedures that the classroom teacher can follow to make maximum use of paper-and-pencil tests as one vehicle for diagnosis. It also outlines some methods for analyzing children's work and provides some suggestions for remediation.

What are the basic facts?

Years ago it was not uncommon to hear teachers exclaim that their pupils did not know their tables! These tables contained the basic number facts that all children were expected to memorize. Difficulties with basic facts have not disappeared with either the modern mathematics reform movement or time. Children still have trouble with their tables!

The 100 basic addition facts are those in which two one-digit numbers are added to form a sum (examples are $3 + 7 = 10$ and $0 + 9 = 9$). Basic multiplication facts are similar: two single-digit factors are multiplied to form a product (9 × 4 = 36 and 2 × 1 = 2). The basic subtraction facts are related to the basic addition facts (9 + 8 = 17 and 17 − 8 = 9), and in a similar fashion, division is related to multiplication (42 ÷ 7 = 6 and 7 × 6 = 42). There are the same number of basic addition, subtraction, and multiplication facts as addition facts, but since division by zero is ruled out, there are only 90 basic division facts.

The mastery of basic facts is essential for effective computation. Although some children may find answers to basic facts by counting or by some type of grouping procedure, we generally expect them to memorize the basic facts. This does not mean, however, that the facts are to be *taught* by memorizing.

What is an algorithm?

An algorithm is a sequence of steps in the solution of a particular type of problem. These steps are always precise and well defined. A computer can be programmed to calculate by giving it the steps to follow, but children should not be taught to use algorithms without understanding the processes. However, at some point we do expect them to memorize or master algorithms.

Paper-and-pencil algorithms form a large part of the computational program in the schools. The answer to the example 52 − 17 = ? can be found by applying a given algorithm. The following progression illustrates one such algorithm.

1. Arrange the minuend, 52, and the subtrahend, 17, vertically so that the ones-place digit of the subtrahend is lined up under the ones-place digit of the minuend:

$$\begin{array}{r} 52 \\ -\,17 \\ \hline ? \end{array}$$

2. Compare the 7 in the ones place of the subtrahend with the 2 in the ones place of the minuend. Since the 7 is greater than the 2, it is not possible to subtract and obtain a positive whole number.

3. Regroup the 52 in the minuend into 4 tens and 12 ones (the 5 in the minuend represents 5 tens or 50 ones and may be regrouped, or renamed, to 4 tens and 10 ones).

4. Subtract the 7 in the subtrahend from the 12 ones in the minuend, writing down the difference, 5. (The basic fact 12 − 7 = 5 is used here.)

5. Compare the 1 in the tens place of the subtrahend with the 4 in the tens place of the minuend. Since 1 can be subtracted from 4 to give a positive whole number, write down the difference, 3.

6. The difference between the minuend and the subtrahend consists of 3 tens and 5 ones, written as 35, which is the answer. A summary of the steps given results in the following familiar form:

$$
\begin{array}{cc}
4 & 12 \\
\cancel{5} & \cancel{2} \\
-1 & 7 \\
\hline
3 & 5
\end{array}
$$

Diagnosing Difficulty

We can diagnose computational difficulties in several ways. Besides paper-and-pencil tests, which will be discussed in detail, there are the techniques of observation and interview, both of which can greatly enhance an effective ongoing program of diagnosis. These person-to-person techniques provide an opportunity to discover behaviors that children's papers cannot reveal. Further, the analysis of written work involves a time lag, whereas on-the-spot diagnosis allows remediation to begin immediately.

Naturally, a teacher will be observing pupils' behavior no matter what other means of diagnosis is employed. The firsthand observation of computational difficulties will obviate longer periods of analyzing written work. Often just the puzzled expression on a child's face can alert the teacher to a problem.

Suggestions for interviewing children are found in many references (e.g., Gray [1966], Krutetskii [1976], Lankford [1972; 1974], and Weaver [1955]). Asking a child to "talk an example through" is very effective. The teacher can ask direct questions concerning particular points of difficulty, and the pupil's attitude and response can be gauged. But interviews, although they give more information than an analysis of a child's written work, are time-consuming; it is difficult for a teacher to find time to have frequent individual interviews with each member of the class. Consequently, it is important to develop varied means of diagnosis, using interviews where other techniques prove ineffective.

The well-developed paper-and-pencil test enables a teacher to diagnose many kinds of computational difficulties. We shall discuss the development and preparation of a suitable test, the administration of the test, and its interpretation.

Preparing the test

Four steps are used in developing a diagnostic test: (1) select the types of computational examples to be evaluated, (2) develop at least three examples for each type, (3) prepare the test for duplication, and (4) make a suitable answer key.

Try to select examples that will facilitate the analysis of error patterns. Using three examples for each type of computational skill will increase the validity of the analysis and aid in pinpointing particular patterns. Using fewer than three may be ineffective in that the pattern tends to be obscured by difficulties with particular basic number facts. (A brief discussion of error-pattern analysis will follow later. For additional help in error-pattern analysis, consult references such as Ashlock [1976], Cox [1975 (a) and (b)], Pincus [1975], West [1971], and the article in this yearbook by Carl Backman.)

Many error patterns are unique to a particular kind of example. Hence, it is important that the types be clearly and, if possible, uniquely described. Also, the sequencing of types may be important, particularly when the diagnostic test is used as a pretest or posttest.

Assume that after working with two-digit multipliers for some time, you find that your pupils are still having trouble. Five types of examples that can be used to pinpoint the source of difficulty are shown in figure 10.1.

Type I: A two-digit factor and a one-digit factor, with regrouping. Examples:

$$\begin{array}{ccc} 26 & 34 & 53 \\ \times\ 5 & \times\ 6 & \times\ 4 \end{array}$$

Type II: A three-digit factor and a one-digit factor, with regrouping from ones to tens and tens to hundreds place. Examples:

$$\begin{array}{ccc} 126 & 134 & 153 \\ \times 5 & \times 6 & \times 4 \end{array}$$

Type III: A three-digit factor (with zero in the tens place) and a one-digit factor, with regrouping from ones to tens place. Examples:

$$\begin{array}{ccc} 206 & 304 & 503 \\ \times 5 & \times 6 & \times 4 \end{array}$$

Type IV: A three-digit factor and a two-digit factor (multiple of ten), with regrouping in the second partial product. Examples:

$$\begin{array}{ccc} 126 & 134 & 153 \\ \times 50 & \times 60 & \times 40 \end{array}$$

Type V: A three-digit factor and a two-digit factor, with regrouping in the second partial product only. Examples:

$$\begin{array}{ccc} 126 & 134 & 153 \\ \times 51 & \times 62 & \times 41 \end{array}$$

Fig. 10.1

You need not write descriptions of the types of examples, but do take care to ensure that each example represents the same level of difficulty. It is also wise (as in the examples given) to reuse certain basic facts for each type. Establishing some control over the use of basic facts will enhance the interpretability of the test. In the particular examples selected, two constraints were exercised. The first was that no facts were used employing 7, 8, or 9 as factors. The second constraint involved limiting the number of basic facts by reusing the same ones in different types of examples.

A simple means for applying these constraints is to build a matrix such as the one illustrated in figure 10.2. The matrix will help you restrict the basic facts to those that are generally known by most of your pupils. The matrix can also be used to reduce the variety of facts. Since you can control the facts used, the most significant variable in your test will be the type of example. It is to these types of examples that we shall direct our analysis.

First factor	Second factor						
	0	1	2	3	4	5	6
0		+++	+	++	+	+	+
1		++	+	+		+	+
2		+		+	+		
3							
4	+	+++		+++ ++		+++ ++	
5	+	+++	+++ ++				+++ ++
6	+	+++		+++ ++	+++ ++		

Fig. 10.2. Analysis of basic facts used in diagnostic test

When you prepare the test for duplication, mix the examples and provide working space on the form. Seeing how a child works an example can be as important to your diagnosis as the answer itself. To make your diagnosis easier, distinguish each type of example in your answer key. For instance, enclose each type in a different shape (e.g., circle, square, triangle, etc.). Figure 10.3 illustrates this kind of answer key (and corresponds to the examples given in fig. 10.1).

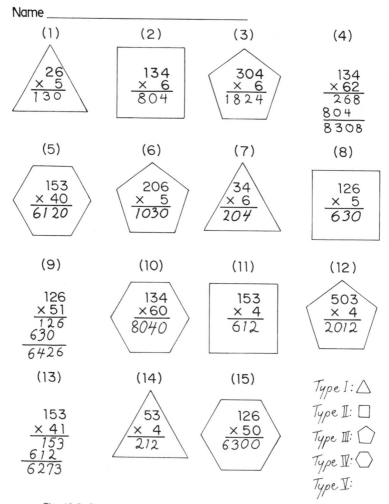

SAMPLE DIAGNOSTIC TEST ANSWER KEY
WITH TYPES IDENTIFIED

Name _____

(1)

$$\begin{array}{r} 26 \\ \times\ 5 \\ \hline 130 \end{array}$$

(2)

$$\begin{array}{r} 134 \\ \times\ 6 \\ \hline 804 \end{array}$$

(3)

$$\begin{array}{r} 304 \\ \times\ 6 \\ \hline 1824 \end{array}$$

(4)

$$\begin{array}{r} 134 \\ \times 62 \\ \hline 268 \\ 804 \\ \hline 8308 \end{array}$$

(5)

$$\begin{array}{r} 153 \\ \times 40 \\ \hline 6120 \end{array}$$

(6)

$$\begin{array}{r} 206 \\ \times\ 5 \\ \hline 1030 \end{array}$$

(7)

$$\begin{array}{r} 34 \\ \times 6 \\ \hline 204 \end{array}$$

(8)

$$\begin{array}{r} 126 \\ \times\ 5 \\ \hline 630 \end{array}$$

(9)

$$\begin{array}{r} 126 \\ \times 51 \\ \hline 126 \\ 630 \\ \hline 6426 \end{array}$$

(10)

$$\begin{array}{r} 134 \\ \times 60 \\ \hline 8040 \end{array}$$

(11)

$$\begin{array}{r} 153 \\ \times\ 4 \\ \hline 612 \end{array}$$

(12)

$$\begin{array}{r} 503 \\ \times\ 4 \\ \hline 2012 \end{array}$$

(13)

$$\begin{array}{r} 153 \\ \times 41 \\ \hline 153 \\ 612 \\ \hline 6273 \end{array}$$

(14)

$$\begin{array}{r} 53 \\ \times\ 4 \\ \hline 212 \end{array}$$

(15)

$$\begin{array}{r} 126 \\ \times 50 \\ \hline 6300 \end{array}$$

Type I: △
Type II: □
Type III: ⬠
Type IV: ⬡
Type V:

Fig. 10.3. Sample answer key to diagnostic test with types identified

Administering the test

The administration of the test (which can be given to one child as well as to a large or a small group) should be planned to obtain as much information as possible. Allow adequate time for all the children to complete the test. Encourage them to show all their work on the test papers; no "scratch paper" should be used. Make it clear that this test is designed to help them and is not for a grade. Observe the children during the test: look for the reluctant or dawdling child, the "counter" (who counts in the air or uses tongue and teeth, the classroom clock, or fingers), the child who looks puzzled, the child who works with an air of confidence.

Interpreting the test

When the test is completed, you are ready to analyze the results. Feel free to make notes on the papers. The tests need not be kept from the pupils and may indeed prove helpful in remediation, but their major use will be for your diagnosis. You may wish to keep them on file, especially the tests of those children you know are having difficulties.

A systematic approach will facilitate your interpretation of the test results. The first step is to *decide how you will interpret each number of missed examples* for each type on the test. It is reasonable to assume that a child who has missed none of the examples for a given type knows how to do that type. However, correct answers are not absolute insurance that the child understands. There are times when a pupil obtains correct answers for all the examples but has used a faulty procedure. This happens infrequently, but it is pointed out to emphasize the need for careful analysis. An example of this will be given when error patterns are discussed. It is also true that a child who does not understand what he or she is doing might still compute accurately using rote memory. Note that although correct answers are not absolute insurance that the child understands, the major concern will be with those children who do not do the examples correctly.

Some errors are careless (or random). A useful criterion to follow is that one incorrect answer from three examples may be considered a careless mistake unless there are indications to the contrary.

When two examples are missed, there is more information to process; the possibility of recognizing an error pattern, if it exists, is increased. Assume that a child who misses two out of three does not know how to do that particular type of example. This assumption means that you will be looking for the *cause* of the difficulty. Subsequently, you may find that both were careless errors. Or the single correct example may indicate the effect of compensating errors, chance, or the observation of another child's paper.

Three incorrect responses indicate that you should make a careful analysis of the errors that have been made. But what is to be done when a child does not respond to any (or some) of the examples? Many teachers consider no answer an incorrect response. No answer is certainly undesirable, but it is neither correct nor incorrect. Remediation for no response may be quite different from that for incorrect responses.

Having decided what you will do about each number of errors for each type, the second step (and most obvious!) is to *note the incorrect answers.* Draw attention to them with a colored pencil. Do not merely mark an example wrong but circle the incorrect part or parts of the answer. This is important. Look for regrouping errors, incorrect operations, apparent basic-fact errors, and similar items that appear to contribute to the error. The use of some sort of shorthand notation may be helpful at this point. Also draw attention to marks on the paper that indicate the child has counted or is making tallies to get the answers.

The third step is to *make an analysis of the child's work on the test.* You may discover that a child (1) demonstrates an error pattern, (2) makes random errors with no pattern, (3) knows how to do the algorithm but does not know the basic facts, or (4) demonstrates combinations of these three types of errors.

Error patterns may be recognized by analyzing the child's methods of computation. However, if a child has missed some of the basic facts, the pattern will be more difficult to determine. Consequently, it is important that you combine information from a *basic-facts test* with that obtained from the diagnostic test.

Since a basic-facts test provides an important dimension in your diagnosis, you may wish to make one up and use it *prior* to the administration of the diagnostic test. Then you will have information regarding basic-fact errors for every child who takes the diagnostic test. Or, if you prefer, the facts test may be given *after* the diagnostic test and then only to those children from whom the basic-facts information is desired.

Test only the specific facts that will be needed on the diagnostic test. Time the test to identify those who have mastered the facts. The difference between commutative pairs, such as 8×6 and 6×8, may be disregarded to shorten the test. For the diagnostic test illustrated in figure 10.3, only twenty-six basic facts are needed. Figure 10.4 gives an example of a basic-facts test that incorporates the suggestions above and is designed to be used with the sample diagnostic test. (See also fig. 10.2.)

(1) $1 \times 1 =$	(10) $4 \times 3 =$	(19) $6 \times 3 =$
(2) $6 \times 4 =$	(11) $6 \times 0 =$	(20) $4 \times 0 =$
(3) $5 \times 0 =$	(12) $2 \times 4 =$	(21) $5 \times 2 =$
(4) $0 \times 2 =$	(13) $1 \times 2 =$	(22) $0 \times 6 =$
(5) $5 \times 1 =$	(14) $0 \times 1 =$	(23) $1 \times 5 =$
(6) $1 \times 3 =$	(15) $4 \times 5 =$	(24) $0 \times 3 =$
(7) $5 \times 6 =$	(16) $1 \times 6 =$	(25) $2 \times 1 =$
(8) $6 \times 1 =$	(17) $2 \times 3 =$	(26) $4 \times 1 =$
(9) $0 \times 5 =$	(18) $0 \times 4 =$	

Fig. 10.4. Sample basic-facts test

Panofsky, Erwin. *Meaning in the Visual Arts*. Garden City, N.Y.: Doubleday & Co., Anchor Books, 1955; Chapter 6.

Porter, A. T. *The Principles of Perspective*. London: University of London Press, 1927.

Richter, Irma. *Rhythmic Form in Art*. London: John Lane, 1932. 127 pp.
Principles of perspective and dynamic symmetry as exemplified in the works of the great masters.

Schudeisky, Albrecht. *Geometrisches Zeichnen*. Leipzig: Teubner, 1919. 99 pp.

Wolff, Georg. *Mathematik und Malerei*. Leipzig: Teubner, 1925. 85 pp. (Paper)
Perspective in painting; composition; brief bibliography.

Wulff-Parchim, L. *Dürer als Mathematiker*. 1928.

12.5 The Golden Measure—Dynamic Symmetry

"The Golden Section therefore imposes itself whenever we want by a new subdivision to make two equal consecutive parts or segments fit into a geometric progression, combining thus the threefold effect of equipartition, succession, continuous proportion; the use of the Golden Section being only a particular case of a more general rule, the recurrence of the same proportions in the elements of a whole."—Heinrich Timerding.

Amata, Sister M. A mathematical secret of beauty. *Summation (ATM)*, vol. 11, no. 6, pp. 50–57; June 1966.

Bax, James A. The Golden Section. In *The Mathematics βαˣ*, (Dept. of Math., Florida Atlantic University), vol. 2, no. 1, pp. 1–8; Mar. 1967.

Beard, Robert S. Powers of the Golden Section. *Fib.Q.* 4:163–67; Apr. 1966.

Beiler, Albert H. *Recreations in the Theory of Numbers*. Dover, 1964.
"Theorema Aureum," pp. 200–210; Golden Section.

Berg, Murray. Phi, the Golden Ratio (to 4599 Decimal Places), and Fibonacci Numbers. *Fib.Q.* 4:157–62; Apr. 1966.

Borissavlievitch, M. *The Golden Number and the Scientific Aesthetics of Architecture*. New York: Philosophical Library, 1958. 96 pp.

Brooke, Maxey. The Section called Golden. *J.R.M.* 2:61–64; Jan. 1969.

Coxeter, H. S. M. *Introduction to Geometry*. New York: John Wiley & Son, 1961.
Chapter 11: "The Golden Section and Phyllotaxis." Deals with mean and extreme ratio; *de divina proportione*; the golden spiral; the Fibonacci numbers; phyllotaxis.

Frisinger and Dence. Problem 177. *Pentagon*, vol. 24, no. 2, pp. 102–3; Spring 1965.
Note on dynamic symmetry.

where the sequence of steps and prerequisite skills are meticulously reconstructed.

Some of the difficulties that can be diagnosed using the techniques developed earlier are these:

1. Errors that have an identifiable pattern and are logically consistent in the pupil's way of thinking
2. Random errors
3. Difficulty with basic facts
4. Failure to respond
5. Counting, either with markers on the test paper or with objects (e.g., fingers, clock numerals)

A child may have more than one of these difficulties, but for our purposes, remediation will be discussed separately for each.

Error patterns

Several approaches are helpful in dealing with error patterns: (1) develop and examine the sequence of steps leading up to the algorithm in which the error pattern occurs, (2) provide physical models for manipulative activity, (3) develop a diagram approach to the algorithm, and (4) use an alternative algorithm.

For each algorithm there is a prior sequence of learning steps. Analyzing such a sequence will allow you to discover the point at which the child has gotten into trouble. Starting with the examples that have given difficulty, work backward, noting which steps would lead up to this form of the algorithm. It is possible that the child was introduced to the particular algorithm as a short form for a longer, more elaborate one. The transition from the longer form to the short form should be considered a point of possible difficulty. Developing a sequence of types of examples (as you did in preparing the diagnostic test) may also be used as a criterion for gauging the point of difficulty. Look for problems where the transition from one type of example to another calls for renaming or regrouping. Sometimes the notation used will be the only difficulty. Also look for problems with the troublesome zero and confusion over how to record aids in computation ("crutches"). The child may be taken through a sequence to reaffirm your initial diagnosis. Reteaching the sequence will then prove useful for remediation.

An activity related to sequencing is the use of diagrams or flowcharts. Diagrams representing the steps for the algorithm aid in directing the child's attention to the process. You may wish to introduce additional steps in the algorithm to portray the process graphically. Breaking the algorithm down into small segments may help. Using a flowchart may also be in order as a logical method showing how the algorithm is processed (see Kessler [1970]).

Providing models may be appropriate for work with addition, subtraction, and the simpler forms of multiplication and division. Just working the problem with physical materials may be helpful in showing the child where an error exists. Some of the models that can be used are (1) place-value and pocket charts; (2) sticks or slips of paper for bundling and developing ideas of place value; (3) beans and bean sticks, where beans are affixed to tongue depressors to represent tens; (4) base-ten blocks, which are illustrated by the Dienes multibase blocks; (5) the open or closed abacus; (6) chip trading, where different-colored chips represent ones, tens, hundreds, and so on; and (7) counters of various sorts, which can be grouped and separated into piles of ten and powers of ten.

There are times when substituting an alternative algorithm for the one the child is using may be appropriate. One of these times is when you want to break an established pattern of failure and give success, even at the expense of teaching a mechanical process (see Ashlock [1976, pp. 11–12] for a discussion of this alternative). Using an alternative algorithm may involve the development of some skills and ideas that were not needed for the process the child was originally taught. Using another algorithm may compel the pupil to look at the process from a fresh point of view. This procedure is probably the one to take when all else fails. In addition to the traditional alternative algorithms, recent work with "low stress" algorithms shows promise for this approach to remediation (see Hutchings [1975; 1976]).

Random errors

If the errors you find are random ones, little can be done by way of systematic remediation. However, some children make more random errors than they should. Carelessness is one cause of this type of error. Marie may add when the example calls for subtraction. Devising simple exercises that force her to focus on the operation sign may help. For instance, have her circle the operation sign before beginning the algorithm.

Messiness and poorly organized work will also contribute to random errors. If Mike has a tendency to make mistakes because he fails to make straight columns, providing graph paper may be one intermediate solution. Having him turn lined paper sideways is also helpful in keeping columns organized. Specific help in forming numerals may aid some children, and a positive reward for neatness always helps. Displaying neat papers is one type of reward; giving praise is another.

Some random errors, of course, just happen! Hardly anyone is free of this sort of error. Having reasonable levels of expectation for children's papers may help to reduce the onus attached to random error. Expecting

100 percent accuracy for every test every time may be too severe. Accepting error without condoning it is a good attitude for a teacher to develop. It will go a long way toward helping a child eliminate errors without becoming overly obsessed with total accuracy.

Difficulty with basic facts

There are two types of difficulties with basic facts. The first occurs when the child does not understand the concept of the operation. For instance, not knowing what multiplication is will lead to basic-fact difficulty even when the child has some of the facts memorized. This type of difficulty, although not easy to ascertain, may be discovered as you observe the child working. It is not likely that a paper-and-pencil test will reveal it. You must use some subjective judgment in making this assessment. Having done so, you will need to go back to the most basic ideas. Models can be used; diagrams are helpful; the manipulation of objects should certainly be employed. Certain forms of learning disability may warrant special help from a trained clinician.

The second and more common type of basic-fact difficulty is the child's failure to memorize the facts. Betsy may understand the process, but she may not have had enough repetition. Practice is called for with this diagnosis. Such practice or drill may take many forms and *can* be interesting. Among the general suggestions for developing good drill material, consider the following: (1) provide for variety, using cassette recordings, flash cards, games, and group recitations as variants on the worksheet method; (2) provide immediate feedback for every exercise designed for practice, including homework; (3) carefully plan the drill exercises, choosing appropriate examples and spacing the practice for maximum effect without boredom; and (4) always evaluate students' learning of the basic facts using a timed test. For a more thorough discussion of principles for drill, see the article by Edward Davis in this yearbook.

Failure to respond

The child who does not respond on a diagnostic test may or may not know how to work the examples. The teacher should take time to investigate and try to pinpoint the difficulty. An interview is appropriate; just asking a child to calculate an example for you will often lead to the source of the problem. Karen may know how to do the work, but she may be overly cautious and fearful of making a mistake. Jeff may not know how and *know* that he doesn't! Dan's failure to respond may result from some sort of personality difficulty.

In one respect, Karen's problem is easier to deal with than the error-pattern type. At least she has not developed her own original algorithm!

(No need to "erase" here.) When the problem seems to be a lack of confidence, games may help, since they encourage self-competition. It may take some searching to find the right game to match a child's difficulty; the game should be interesting and appealing as well as appropriate to the child's particular computational needs. For some suggestions, see the article by Robert Ashlock and Carolynn Washbon in this yearbook. Another aid to building confidence in a child is cross-age tutoring or peer help.

Those pupils, such as Jeff, who know that they don't know how to do the work will demonstrate this fact in an interview. This is one of the easier remedial situations. If only more of your pupils would let you know in this manner rather than always putting something down on their papers! Jeff does not seem to be afraid to let you see that he doesn't know something, and if he wants to learn it, the remedial approach can be direct and simple.

Coping with Dan's problem may be the hardest of all. Most teachers are not in a position to give therapy, but they can all accept children for what they are and what they feel. Accepting children's work (or lack of it) and seeking to encourage them may be the best remediation.

Counting

How often have you seen a child using fingers to count? When was the last time you observed hundreds of little "rabbit tracks" on the back of a multiplication paper? These are common types of behavior and indicate that you have a "counter" on your hands. Some teachers treat counters as offenders! However, it will be difficult to identify the counter who is getting correct answers as an error-maker. Counting is inefficient and frequently leads to errors. Nevertheless, a good counter probably has greater understanding than most of the children who have error patterns. Negative feedback seldom cures counters. They resort to covert methods that tend to defeat diagnosis.

One approach to take with the counter is to speed up the process to the point where the child recognizes the need to memorize. Make a game of timed tests, starting with one that can be completed using the counting process. Then make the tests increasingly more demanding in terms of time needed. Another effective technique is to involve the counter in a game of basic facts where the first response wins. Or use a game in which some physical activity is associated with the response to a basic fact. For example, call out the fact, toss a beanbag to the child, and award points for answers given before the bag is caught. The problem of the counter is not always easy: it requires persistence and creativity on the part of the teacher. With some children it is only necessary to suggest that *older* boys and girls do not count.

Conclusion

Diagnosis and remediation are intimately related. Without good diagnosis, the best remediation can only be a shot in the dark. Without remediation, diagnosis is an exercise in frustration. Remediation follows diagnosis and gives validity to it. Good diagnosis enables the teacher to concentrate on a child's particular problem. Using the techniques suggested here should help the classroom teacher to make more effective diagnoses. The suggestions for remediation are of a general nature. However, each teacher will find that accurate diagnosis often suggests a cure. The diagnostic process also permits the teacher to use skill and insight in meeting individual needs.

Let your diagnoses be spurs to direct you to remedial measures. As your skill in diagnosis increases, so should your ability to give remediation. With effective remedial action following accurate diagnosis, you have an unbeatable combination in dealing with pupils' computational difficulties!

REFERENCES

Ashlock, Robert B. *Error Patterns in Computation: A Semi-Programmed Approach.* 2d ed. Columbus, Ohio: Charles E. Merrill Publishing Co., 1976.

Cox, L. S. (a). "Diagnosing and Remediating Systematic Errors in Addition and Subtraction Computations." *Arithmetic Teacher* 22 (February 1975): 151–57.

———(b). "Systematic Errors in the Four Vertical Algorithms in Normal and Handicapped Populations." *Journal for Research in Mathematics Education* 6 (November 1975): 202–20.

Gray, Roland F. "An Approach to Evaluating Arithmetic Understandings." *Arithmetic Teacher* 13 (March 1966): 187–91.

Hutchings, Barton. "Low-Stress Algorithms." In *Measurement in School Mathematics,* 1976 Yearbook of the National Council of Teachers of Mathematics. Reston, Va.: The Council, 1976.

———."Low-Stress Subtraction." *Arithmetic Teacher* 22 (March 1975): 226–32.

Kessler, Bernard M. "A Discovery Approach to the Introduction of Flowcharting in the Elementary Grades." *Arithmetic Teacher* 17 (March 1970): 220–24.

Krutetskii, V. A. *The Psychology of Mathematical Abilities in Schoolchildren,* chap. 7. Edited by Jeremy Kilpatrick and Izaak Wirszup. Translated by Joan Teller. Chicago: University of Chicago Press, 1976.

Lankford, Francis G., Jr. *Some Computational Strategies of Seventh Grade Pupils.* Charlottesville, Va.: School of Education, University of Virginia, 1972. (ERIC no. ED 069 496)

———. "What Can a Teacher Learn about a Pupil's Thinking through Oral Interviews?" *Arithmetic Teacher* 21 (January 1974): 26–32.

Pincus, Morris, Margaret Coonan, Harold Glasser, Lillian Levy, Frances Morgenstern, and Herbert Shapiro. "If You Don't Know How Children Think, How Can You Help Them?" *Arithmetic Teacher* 22 (November 1975): 580–85.

Weaver, J. Fred. "Big Dividends from Little Interviews." *Arithmetic Teacher* 2 (April 1955): 40–47.

West, Tommie A. "Diagnosing Pupil Errors: Looking for Patterns." *Arithmetic Teacher* 18 (November 1971): 467–69.

11

Analyzing Children's Work Procedures

Carl A. Backman

DIAGNOSIS in the mathematics classroom is based on obtaining and interpreting evidence of student work procedures through direct observation of students at work, oral interviews, and students' written work. The goal is to design future instruction based on an assessment of children's strengths and weaknesses relative to such factors as their current work procedures, content background, and cognitive level. In using a diagnostic approach to teaching, one usually moves repeatedly through the following series of steps (based on the model of the "diagnostic teaching cycle" presented in Reisman [1977, pp. 1-3]):

1. Identify students' strengths and weaknesses.
2. Hypothesize reasons for the strengths and weaknesses.
3. Set instructional objectives.
4. Determine and implement instructional procedures.
5. Evaluate progress continuously.

The central focus of this article is an examination of a variety of student work procedures and answer patterns for whole-number computation. This examination will lead to a more systematic connection of steps 1 and 2 with steps 3 and 4 in the diagnostic teaching of computational skills.

In planning for instruction on computational skills, the teacher usually must cope with two different types of pupils. Some pupils have had no prior instruction on the skills to be acquired. Other children have already acquired, in one way or another, methods of working with the material to be learned, although some of these methods may be inappropriate.

Although appropriate instructional sequences for the two types of pupils described above may be different, instructional decision-making can be enhanced for both types by examining the different kinds of learning involved in acquiring computational skills and the types of complications usually associated with such learning. These two factors—the type of learning desired and the type of complication expected—form a basis for classifying student work procedures and answer patterns.

Cognitive learning is a complex phenomenon, with many theorists suggesting that there is more than one kind of learning. This idea is readily apparent in the learning of computational skills. The acquisition of such skills frequently involves applying concepts and principles, memorizing information, sequencing a set of steps in the right order, and selecting correct procedures for a given situation. Related to these kinds of learning are such complications as forgetting, associating an inappropriate response with a given stimulus, performing the steps of a procedure in the wrong order, omitting some of the steps of a procedure, confusing elements of one procedure with those of another, using a concept where it is inappropriate, and failing to use a concept where it should apply. Furthermore, written computation is subject to a variety of recording errors.

Teaching children who have had no prior instruction on the computational skills or concepts to be acquired and who do not already have a procedure for doing the computation should involve instructional decisions centering on (a) maximizing the conditions for initial learning and (b) minimizing the conditions that might produce complications. To do this, a teacher analyzes the types of learning to be acquired and the appropriate conditions for their acquisition and also considers the common learning complications. Knowing what types of errors pupils are likely to make allows the teacher to design instruction to prevent errors before they occur.

Teaching pupils who have acquired inappropriate methods for doing computation should involve instructional decisions centering on (a) determining the pupils' present methods of work, (b) eliminating the inappropriate methods or concepts, and (c) developing appropriate computational procedures. Careful examination of student work procedures and the answer patterns they produce can provide guidance for the kind of instruction needed.

The illustrations presented in the section "Student Work Procedures and Answer Patterns" are grouped in eight major categories:

1. Correct answer from a standard procedure
2. Correct answer from a nonstandard procedure
3. No response

4. Random errors
5. Errors related to *conceptual* learning
6. Errors related to *sequencing* steps within procedures
7. Errors related to *selecting* information or procedures
8. Errors related to *recording* work

For each major category, a general description is given for the work procedures and answer patterns in that group. For categories based on the major kinds of learning involved in computational skills, the principal conditions thought to be appropriate for that kind of learning are outlined.

Each of these categories is further analyzed into subgroups related to the types of complications associated with computational skills. *Examples* of student work are given with a brief description of the process the child used to obtain the answer. (The prototypes for the examples were compiled from several sources, including Ashlock [1976], Glennon and Wilson [1972], Mee and Bishop [1975], Pincus et al. [1975], and Reisman [1972].) Often the examples are followed by a set of *notes* suggesting instructional procedures and materials that might be used to eliminate or prevent the error.

A summary of the major groupings and their subcategories is contained in table 11.1.

Some cautions

Before considering the descriptions of work procedures and answer patterns, the specific illustrations of student work, and the instructional suggestions found in the next section, a few cautions are in order.

1. The illustrations of student work by no means exhaust the possible response patterns used in computational exercises.

2. The illustrations of instructional procedures are also by no means exhaustive. They are merely suggestive of the wide variety of possibilities. Many other suggestions can be found in other articles in this yearbook.

3. It is highly probable that there is a degree of overlap between some of the categories in the classification system.

4. The easiest evidence to obtain is student written work. However, interpretation can be difficult at times, since identical written work can be obtained from different work procedures. (For example, the placement of the 6 in the work shown could be an error of digit alignment. It could also indicate a faulty concept of division, a con-

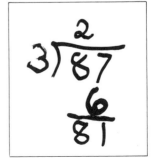

fusion of the procedure with that for multiplication, or a confusion of steps within the division algorithm.) Inasmuch as possible, it is wise to couple an analysis of written work with direct observation of pupils and oral interviews in which children explain what they are doing.

5. For some pupils, it is possible that several different types of errors have been made within a single exercise set.

6. Sometimes consistent but incorrect work procedures will generate correct answers for a child part of the time. This frequently makes it more difficult to convince the child that the procedure is inappropriate.

7. In spite of the fact that most of the illustrations deal with an analysis of pupil errors, the teacher should continuously assess strengths (correct procedures) as well as weaknesses in order to ascertain the appropriate starting point for instruction.

8. At times, an analysis of work procedures alone is insufficient to determine an appropriate course of instruction. Sometimes psychological diagnosis (for motivational or emotional difficulties) or physical diagnosis (for health, sight, or hearing deficiencies) may be required.

9. In focusing on specific work patterns, the teacher may lose track of the sequential nature of much of the development of computational skills. A hierarchy of learning objectives, as typified by contemporary diagnostic tests, textbook scope-and-sequence charts, and school district mathematics skills continua, is an important tool for determining the appropriate content level at which a pupil should be working. Diagnosis will reveal the places in the sequence at which the pupil is currently working successfully. Instruction can then be designed to fill in gaps and to move forward in the sequence.

Student Work Procedures and Answer Patterns

1. Correct answer from a standard procedure

Pupils obtain the correct answer to a computational exercise using one or more of the traditional procedures (or their alternatives) accepted as "standard" in contemporary mathematics curricula.

In performing computations correctly, pupils have very likely—

1. selected a correct procedure;
2. performed the steps of the procedure in the correct sequence;
3. used correct arithmetic facts;
4. applied underlying structural concepts;
5. recorded work appropriately.

Examples:

1.   887   Pupil has used the
      + 926 standard addition al-
              gorithm.
      1813

2.
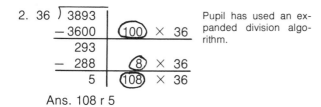
Pupil has used an expanded division algorithm.

Ans. 108 r 5

Notes: Since contemporary mathematics programs differ somewhat in accepted procedures, teachers should be aware of the range of possible procedures their pupils might be using. It is not fair to have an alternative procedure carefully learned from another teacher labeled incorrect.

2. Correct answer from a nonstandard procedure

Students obtain the correct answer to a computational exercise using a procedure not accepted as standard in contemporary mathematics curricula. Frequently, these procedures are devised by the pupils themselves.

Examples:

1.   583   Pupil worked from left to right finding
      −262 differences for each place and then add-
       300   ing the differences.
        20
         1
       321

2.   860   860   Pupil multiplied the greater factor by half
   ×6 is done   ×3   of the other factor, then doubled this
           2580   product.
            ×2
         5160

Notes: Technically, these are not errors. However, the children may be using an immature or inefficient algorithm while actually being capable of using a more efficient one. Care should be taken not to identify their procedures as "wrong," particularly when they can justify what they are doing. If having pupils move to a more efficient procedure is desirable, carefully point out the reasons for wanting them to change (for example, "The new way will take less time," "Your way may cause you to get a different kind of problem wrong," or "Most people do this a different way"). Look for ways to link the children's present procedures to new ones. For example 1 above, the pupil might first be shown how to use the procedure working from right to left, pointing out that subtraction computation is usually done from right to left. The child can then be shown how to save some time by recording each digit of the difference directly in its proper place without having to record all the zero digits.

Original Procedure	First Modification	Second Modification
583	583	583
− 262	− 262	− 262
300	1	321
20	20	
1	300	
321	321	

3. No response

Pupils give no answer to a computational exercise.

Notes: Such errors frequently are due to—

1. insufficient time for the child to complete a set of computational exercises;
2. pupils' recognition that they have never been able to do that type of exercise;
3. pupils' forgetting how to do the exercise.

Sometimes children who do not respond can actually do the computational exercise, but they apparently lack motivation; for some reason they do not want to try the exercise even though they may be capable of doing it correctly. Attempting to reduce possible frustration by allowing more time may help some. For others, cuing by using related but less complex exercises is appropriate. For example, if pupils say that they cannot do an exercise like the one at the left, suggest a simpler exercise like the one to the right:

$$58374 \qquad\qquad 374$$
$$-\ 23281 \qquad\qquad -\ 281$$

For those who truly cannot do the exercise, find the place in the computational skills sequence at which they are successful and then teach from there toward the desired skill.

4. Random errors

Generally, random errors are of two types:

1. Pupils do one or two exercises of a set incorrectly, but the rest of the exercises of that same type are correct.

Notes: Simply showing a child such an error usually elicits the remark, "Did I do that? How dumb!" Occasional errors like this are not cause for alarm. Repeated random errors over extended periods of time may indicate that the pupil is bothered by internal or external distractions not related to mathematics.

2. Pupils do many exercises incorrectly, but there appears to be no consistency to the responses. A pattern may exist, but it is not detected.

Notes: Some children, although not really knowing how to do the computational exercise, may feel that an answer must be produced. They then use whatever process "comes to mind." This may be a legitimate algorithm for use in a different situation or one that they have "created" from elements of several different algorithms. Having them tell you how they do the computation may give an indication of the specific details of the exercise on which they are concentrating and allow you to detect a pattern. For pupils for whom the random errors indicate no real knowledge of the computational procedure, find the place in the computational skills instructional sequence at which the pupil is successful and then teach from there toward the desired skill. In any event, it is not wise to create an expectation among children that an answer must always be produced, thus forcing incorrect responses.

5. Errors related to conceptual learning

Pupils obtain an incorrect answer to a computational exercise because of a deficiency in an underlying concept or principle. Such deficiencies include—

1. the complete absence of a concept or principle;
2. confusing elements of one concept or principle with those of another;
3. recognizing some but not all instances of the concept or principle;
4. recognizing all instances of a concept or principle but also including instances to which the concept or principle does not apply;
5. building concepts or principles on inaccurate information.

Conceptual learning usually involves the grouping and labeling of elements or situations that are related in some systematic way. Once conceptual learning has occurred, a student will be able to identify instances of the concept or principle even though the specific instance was not included in the set of illustrations used to establish the concept or principle. Instructional procedures should—

1. provide a sufficient number of concrete examples of the concept or principle to enable the full range of the concept or principle to be comprehended;

2. provide a sufficient number of examples not included in the concept or principle (i.e., nonexamples) to ensure that the concept or principle will not be extended beyond its proper limit;

3. emphasize those features of situations that make them examples of the concept;

4. use vocabulary and relations already clearly understood by pupils when concepts or principles are to be defined by verbal descriptions.

Errors related to conceptual learning have been subgrouped into the three categories described below.

A. *Errors related to the meaning or properties of an operation.* Pupils generate incorrect basic facts because their conceptualization of an operation is deficient in some systematic way.

Examples:

1.
$$\begin{array}{r} 72 \\ +86 \\ \hline 147 \end{array}$$
Pupil generates basic facts that are consistently one less than required.

2.
$$\begin{array}{r} 39 \\ -22 \\ \hline 16 \end{array}$$
Pupil has written an incorrect basic fact $(9 - 2 = 6)$.

3. $6 \times 0 = 6$

$$\begin{array}{r} 402 \\ \times 6 \\ \hline 2472 \end{array}$$
Pupil believes that the product of a number and zero is the number.

Notes: One could easily point out to pupils that they are wrong, give them the correct basic fact, and provide a correct procedure for generating the facts (such as counting or grouping objects, making tally marks, etc.). However, in order to eliminate the error completely, it is usually necessary to know the specific procedure used by the children to generate their answers. The difficulties in their procedures should be corrected so that they will not occur again. In example 1 above, a child may be using a number line to produce incorrect basic "facts" by counting dots (fig. 11.1) instead of spaces (fig. 11.2). Unless the error in this procedure is noted and a correct procedure substituted, the child may continue to produce incorrect "facts."

Some conceptual errors are peculiar to division exercises involving remainders. Pupils obtain an incorrect answer because they either ignore

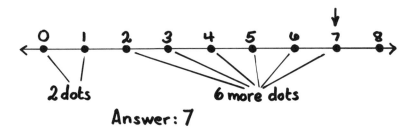

Answer: 7

Fig. 11.1. Original procedure

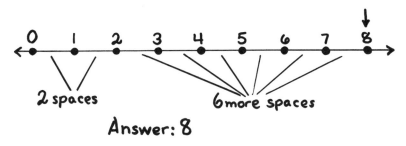

Answer: 8

Fig. 11.2. Modification

final remainders or leave final remainders greater than or equal to the divisor.

Examples:

1.
$$\begin{array}{r} 80 \\ 7\overline{)562} \\ \underline{560} \\ 2 \end{array}$$
Pupil ignored the remainder.

Ans. 80

2.
$$\begin{array}{r} 14 \text{ r } 8 \\ 4\overline{)64} \\ \underline{4} \\ 24 \\ \underline{16} \\ 8 \end{array}$$
Pupil left the remainder greater than the divisor.

Notes: Having children separate sets of objects into subsets corresponding to the exercise should reveal to them that the remainder can be divided further. Using a multiplication check for division computations will reveal an error in example 1, but it will not reveal the error in example 2.

B. *Errors related to the structure of the numeration system.* Pupils obtain an incorrect answer to a computational exercise because they ignore essential features of decimal numerals or have deficient place-value concepts.

Examples:

1.   53   Pupil ignored place value   2.   256   Pupil misaligned addition
    + 4   and added all digits as if   × 39   portion because place-
    ———   they were in the ones place.   ————   value shifts were ignored
     12   2304   when multiplying. (This is
                                                 768   prevalent with zero digits
                                                ————   in the multiplier.)
                                                3072

Notes: In general with such errors, check to see if the children can identify the place values of numerals in the exercise. If they cannot, basic instruction on place-value concepts is needed. To convince them that there is an error, have *the pupils themselves* estimate the answer, check the answer, or construct a simulation using objects or other manipulative aids.

For example 1, try displaying the exercise on a place-value device such as the place-value chart or abacus illustrated in figures 11.3 and 11.4.

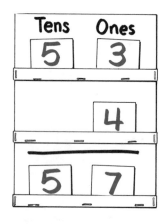

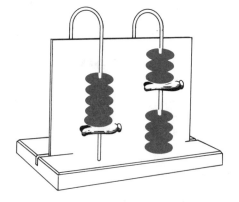

Fig. 11.3   Fig. 11.4

For example 2, expanded algorithms that highlight place value, like the one illustrated in figure 11.5, are appropriate for some students.

	Find the two partial products:		Then add partial products:
To do:			
256	256	256	2304
×39	×30	×9	+7680
	7680	2304	9984

Fig. 11.5

Some students may be able to follow arguments based on properties of operations to show the effects of place value:

$$200 \times 30$$
$$(2 \times 100) \times (3 \times 10)$$
$$(2 \times 3) \times (100 \times 10)$$
$$6 \times 1000$$
$$6000$$

C. *Errors related to renaming and regrouping.* Pupils obtain an incorrect answer to a computational exercise because they do not see a need for regrouping or because they regroup inappropriately. Such errors are often related to conceptual difficulties.

Examples:

1.
$$\begin{array}{r} 27 \\ +56 \\ \hline 73 \end{array}$$
Pupil dropped the ten that should have been grouped with the digits in the tens place.

2.
$$\begin{array}{r} 11 \\ 4\overline{)64} \end{array}$$
Pupil ignored the internal remainder (2) and did not rename from tens to ones.

3.
$$\begin{array}{r} 28 \\ \times 6 \\ \hline 1248 \end{array}$$
Pupil wrote both digits of the basic facts products in the final product.

4.
$$\begin{array}{r} {}^{5}\cancel{6}\,11 \\ 705 \\ -356 \\ \hline 259 \end{array}$$
Pupil renamed each time from the extreme left place.

Notes: In addition to the suggestions listed above for errors related to the structure of the numeration system, activities that highlight the "exchange" characteristic of renaming and regrouping may be appropriate. In some of these activities, counting individual objects and regrouping them into sets of ten (and so on) in a way that allows the child to observe the identity of the individual objects within the sets of ten can be used. For example 1, the child could count marbles and bag them into sets of ten (fig. 11.6). Ask questions such as, "Do we have enough single marbles

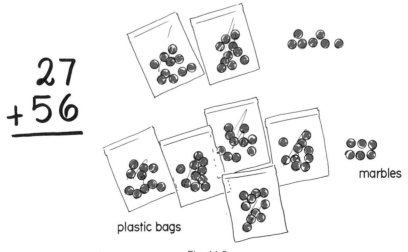

Fig. 11.6

to complete another bag?" Collect ten of the thirteen marbles into a bag and place with the other seven bags. This shows the renaming and regrouping of the thirteen marbles into one ten and three individual pieces.

For some activities, use a grouping procedure that masks the identity of the units when grouped into sets of ten, a hundred, and so on. For example 2, a situation based on an exchange of money can be constructed. Four people are to share $64 given as six ten-dollar bills and four one-dollar bills. Pupils are told that they may exchange ten-dollar bills for one-dollar bills to make the sharing easier. Before any exchange, each person gets one ten-dollar bill, leaving two ten-dollar bills not shared. Each of these is exchanged, giving twenty one-dollar bills to put with the other four one-dollar bills. These twenty-four can be shared by giving each person six one-dollar bills. Thus, each person gets one ten-dollar bill and six one-dollar bills, for a total of $16.

Errors such as the one in example 4 are sometimes the result of faulty teaching when working with addition or subtraction algorithms for two-digit numerals. If pupils are continually instructed to rename the digit on the *left* in exercises like

$$\begin{array}{r} 53 \\ -\ 27 \\ \hline \end{array}$$

they may apply that same instruction when there are more than two digits. It is better to say, "Rename the digit in the place *next to* the ones digit," or, "Rename the digit *immediately* to the left of the ones digit."

6. *Errors related to sequencing steps within procedures*

Pupils obtain an incorrect answer to a computational exercise because they either do the steps of a procedure in the wrong order or leave out some steps. Such errors may be due to memory lapses or confusion among the steps of a procedure. They may also be due to incomplete acquisition of the steps of a procedure or to the acquisition of a faulty procedure.

When teaching for sequencing, take care to see that the children are first able to do each individual step of the procedure. Once skills for each step have been mastered, the procedural sequences can be built. (This is particularly true for the multiplication and division algorithms, which require a certain degree of mastery of the addition and subtraction algorithms.) Procedural sequences should be developed gradually, introducing one new step at a time. Move from practice on single-step procedures to those involving two steps, and so on, to the more complex procedures. Carefully designed repetitive practice is necessary to fix the skills.

Errors related to sequencing have been subgrouped into the two categories described below.

A. *Incorrect order of the steps in a procedure.* Pupils do all the steps of a procedure but in the wrong order. Such reversals may occur in vertical or horizontal steps. Incorrect order in horizontal movement may be caused by interference from the children's familiarity with the normal left-to-right reading order or by confusion with the sequence in an algorithm for another operation.

Examples:

1.
$$\begin{array}{r} 537 \\ -\ 298 \\ \hline 361 \end{array}$$
Pupil subtracted lesser digit from the greater digit regardless of whether it is in the subtrahend or in the minuend.

2.
$$4\overline{\smash{)}418}\quad{}^{142}$$
Pupil reversed role of dividend and divisor in at least part of the division.

3.
$$\begin{array}{r} {}^{2} \\ 48 \\ \times\ 3 \\ \hline 184 \end{array}$$
Pupil added regrouped digit to one factor of the next product and then multiplied.

4.
$$\begin{array}{r} {}^{4} \\ 285 \\ +\ 367 \\ \hline 5116 \end{array}$$
Pupil worked from left to right.

5.
$$\begin{array}{r} 16 \\ 4\overline{\smash{)}244} \\ 24 \\ \hline 4 \\ 4 \\ \hline \end{array}$$
Pupil wrote first quotient digit in extreme right position rather than in the extreme left position.

Notes: Sequencing errors are often related to conceptual difficulties, especially those with renaming and regrouping. At times children will change the order of steps in order to avoid renaming and regrouping. Instruction on renaming procedures may help eliminate this kind of order reversal. Pupils need to be reminded from time to time that the order of the steps in the algorithms for addition, subtraction, and multiplication does not follow the usual order for reading in English. This is not so, however, for division. The division algorithm is basically a left-to-right sequence, although the substeps involving subtraction and multiplication have right-to-left components.

Many children will need examples to follow in which the order of the steps of the computational procedure has been clearly indicated, as in figure 11.7. Giving them a completed computation as an example frequently gives no clues to the first step!

The individual steps of a given computational exercise might be printed on separate index cards and the children asked to put the steps in the correct order. (A similar activity can be used with the incomplete-procedure errors described below by leaving out one of the steps of the procedure and having the children tell which one is missing.)

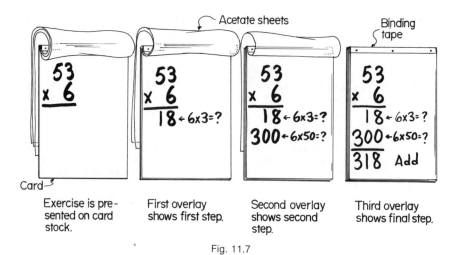

Fig. 11.7

B. *Incomplete procedures.* Pupils obtain an incorrect answer to a computational exercise because they omit a step in a procedure or only partially complete a step.

Examples:

1.
$$\begin{array}{r} 498 \\ -\ 46 \\ \hline 52 \end{array}$$
Pupil found the difference in fewer places than required.

2.
$$\begin{array}{r} 27 \\ \times\ 64 \\ \hline 1228 \end{array}$$
Pupil multiplied the ones digit by the ones digit and the tens digit by the tens digit.

3.
$$\begin{array}{r} 16\ r\ 8 \\ 4\overline{)72} \end{array}$$
Pupil left remainder greater than divisor.

4.
$$\begin{array}{r} 90 \\ 4\overline{)370} \end{array}$$
Pupil stopped at first partial quotient.

5.
$$\begin{array}{r} 3241 \\ 24\overline{)6482} \end{array}$$
Pupil used only the first digit of the divisor to find the quotient.

Notes: For errors of sequence, estimating or checking answers will usually reveal an incorrect answer to the children. Assuming that the children can do each of the individual steps of a given procedure, teachers can render the greatest help in eliminating sequencing errors by providing sufficient practice on exercises designed for each of the different procedural sequences.

7. Errors related to selecting information or procedures

Pupils obtain an incorrect answer to a computational exercise because they use information or procedures appropriate for a different kind of exercise. For one reason or another—memory interference, environmental distractions, visual/perceptual difficulties, and so on—pupils do not notice the cues that differentiate one exercise type from another. (It is assumed here that the children are able to do the correct procedure but that they have selected the wrong procedure for the given exercise.) The pupils need instruction on how to tell when to use one procedure and when to use a different one. The features differentiating one exercise type from another need to be emphasized. This should be followed by practice on sorting exercises into different types.

Errors related to selecting information and procedures have been subgrouped into the four categories described below.

A. *Correct algorithm for the operation, but basic facts for a different operation.* Pupils obtain an incorrect answer to a computational exercise because they associate the stem of a basic fact in one operation with the answer to the similarly constructed fact for another operation.

Examples:

1.
$$\begin{array}{r} 45 \\ + 30 \\ \hline 70 \end{array}$$
Pupil used addition procedure but used multiplication fact $0 \times 5 = 0$ in place of $0 + 5 = 5$.

2.
$$\begin{array}{r} 53 \\ \times 2 \\ \hline 105 \end{array}$$
Pupil used multiplication procedure but used addition fact $2 + 3 = 5$ in place of $2 \times 3 = 6$.

B. *Algorithm and basic facts for a different operation.* Pupils obtain an incorrect answer to a computational exercise because they use a procedure and facts for a different operation. Teachers and pupils usually describe this by saying that they "did not pay attention to the operation sign."

Examples:

1.
$$\begin{array}{r} 56 \\ - 23 \\ \hline 79 \end{array}$$
Pupil used procedure and facts for addition.

2.
$$\begin{array}{r} 53 \\ + 6 \\ \hline 318 \end{array}$$
Pupil used procedure and facts for multiplication.

C. *Correct facts for the operation, but algorithm for a different operation.* Pupils obtain an incorrect answer to a computational exercise because they use elements from an algorithm for a different operation, although they use the correct basic facts for the given operation.

Examples:

1.
$$\begin{array}{r} 76 \\ + 9 \\ \hline 175 \end{array}$$
Pupil used correct addition facts but used multiplication algorithm.

2.
$$\begin{array}{r} 87 \\ 3 \overline{)234} \\ 21 \\ \hline 24 \\ 24 \end{array}$$
Pupil wrote first quotient digit in extreme right position (as in addition, subtraction, and multiplication).

D. *Algorithm for a different type of exercise for the same operation.* Pupils obtain an incorrect answer for a computational exercise because they either use elements from a more complex algorithm or use a simpler algorithm than necessary for the type of exercise.

Examples:

1. $\begin{array}{r} \overset{1}{4}2 \\ +\ 37 \\ \hline 89 \end{array}$ Pupil added one to the tens digit sum even though regrouping was not necessary.

2. $\begin{array}{r} 106 \\ 4\ \overline{)\ 64} \\ 4 \\ \hline \varnothing 4 \\ 24 \\ \hline \end{array}$ Pupil divided remainder digit from first difference by the divisor before "bringing down" the necessary digits.

Notes: Written computational exercises for the different operations are remarkably similar to each other. For example, the vertical arrangement

$$\begin{array}{r} 59 \\ 23 \\ \hline \end{array}$$

is the same whether the exercise is addition, subtraction, or multiplication. Horizontal formats are identical for the four operations: $9 + 3$, $9 - 3$, 9×3, $9 \div 3$. For the same pair of numbers, the child must learn four different responses: 12, 6, 27, and 3. Furthermore, the operation symbols themselves resemble each other. (Compare these pairs of symbols: $+ \times$, $+ \div$, $- \div$.) These resemblances are troublesome for any child who is working quickly through a mixed set of exercises, but particularly so for a child with visual/perceptual difficulties.

To help pupils overcome these difficulties, try specific cuing, such as heavy printing of the operation symbol and the inclusion of word clues, and general cuing, such as the insertion of questions on the exercise pages in special symbols or signs (see fig. 11.8).

There is definite need for repetitive practice on exercise sets having structures like the following: set A, all exercises of type 1; set B, all exercises of type 2; set C, mixed exercises of types 1 and 2; set D, all exercises of type 3; set E, mixed exercises of types 1, 2, and 3; and so on.

8. Errors related to recording work

Pupils obtain an incorrect answer to a computational exercise because of the way in which they record work as opposed to confusion or a lack of knowledge of concepts or computational procedures.

Errors related to recording work have been subgrouped into the three categories described below.

A. *Incorrect numeral formation.*

Examples:

1. $\begin{array}{r} 23 \\ +\ 34 \\ \hline 51 \end{array}$ Pupil read answer as "fifty-one."

2. $\begin{array}{r} 47 \\ -\ 12 \\ \hline \text{ƐƧ} \end{array}$ Pupil wrote digits backwards.

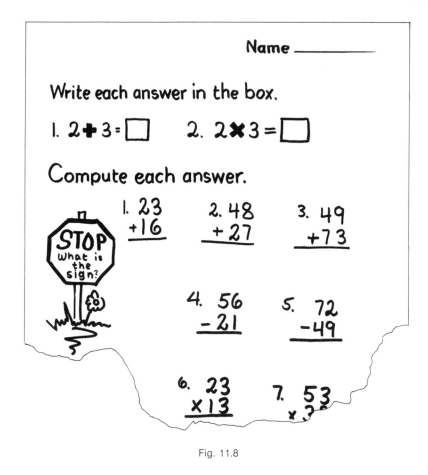

Fig. 11.8

Notes: Provide adequate time for children to complete exercises—frustration due to time pressure can contribute to penmanship difficulties. Provide ample opportunities to record work carefully and accurately and to practice the formation of numerals. Guiding youngsters' hands and numeral-tracing activities with the stroke directions and starting points clearly indicated may also help.

B. *Digit transposition.* Pupils obtain an incorrect answer to a computational exercise because they write digits in the wrong order.

Examples:

```
               3
1.      8      98        Pupil transposed the digits in one
      + 5     + 45       basic fact and regrouped using
       31      161       the wrong digit.
```

Notes: Digit transposition often occurs with basic facts producing sums in the teens, since the oral name for most of the teen numbers names the ones digit first. Asking children to write the name of the number in question to the side of the written work for the exercise will usually cause them to find and eliminate the error on their own.

C. *Misalignment of digits.*

Examples:

1.
$$\begin{array}{r} 385 \\ \times\ 237 \\ \hline 2695 \\ 1155 \\ 770 \\ \hline 888{,}195 \end{array}$$
Pupil added digits in the wrong columns.

2.
$$\begin{array}{r} 426 \\ -\ 21 \\ \hline 216 \end{array}$$
Pupil aligned places from the extreme left rather than from the extreme right.

Notes: Alignment problems can often be avoided or eliminated by providing paper with vertical lining. Try using large-squared graph paper or notebook paper rotated 90°, as shown in figure 11.9.

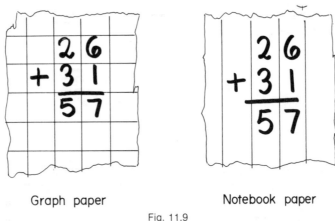

Graph paper   Notebook paper

Fig. 11.9

Summary

Table 11.1 provides a summary of the classification system developed in this article for categorizing student work procedures and answer patterns. You will probably wish to analyze the work of your pupils so you can add other examples to this collection.

Many instructional suggestions for working with children who make errors are included here and elsewhere in this yearbook. As you find additional instructional activities that work well in helping children master computational skills and help to overcome the difficulties described in this article, build a file of these activities for yourself and share them with other teachers. Your teaching of computational skills will acquire greater precision if you keep in mind when setting objectives and planning instruction the nature of the skills to be acquired, the errors pupils are likely to make or have made, and the appropriate instructional activities related to each of these.

TABLE 11.1
SUMMARY OF CATEGORIES FOR
STUDENT WORK PROCEDURES AND ANSWER PATTERNS

1. Correct answer from a standard procedure
2. Correct answer from a nonstandard procedure
3. No response
4. Random errors
5. Errors related to conceptual learning
   A. Meaning and properties of an operation
   B. Structure of the numeration system
   C. Renaming and regrouping
6. Errors related to sequencing steps within procedures
   A. Incorrect order of the steps in a procedure
   B. Incomplete procedures
7. Errors related to selecting information or procedures
   A. Correct algorithm for the operation, but some basic facts for a different operation
   B. Algorithm and basic facts for a different operation
   C. Correct facts for the operation, but an algorithm for a different operation
   D. Algorithm for a different type of exercise for the same operation
8. Errors related to recording work
   A. Incorrect numeral formation
   B. Digit transposition
   C. Misalignment of digits

REFERENCES

Ashlock, Robert B. *Error Patterns in Computation.* 2d ed. Columbus, Ohio: Charles E. Merrill Publishing Co., 1976.

Glennon, Vincent J., and John W. Wilson. "Diagnostic-Prescriptive Teaching." In *The Slow Learner in Mathematics,* Thirty-fifth Yearbook of the National Council of Teachers of Mathematics, pp. 282–318. Reston, Va.: The Council, 1972.

Mee, Roger, and John Bishop. *Teacher's Guide for Individually Diagnosed Error Analysis System.* Columbus, Ohio: Charles E. Merrill Publishing Co., 1975.

Pincus, Morris, Margaret Coonan, Harold Glasser, Lillian Levy, Frances Morgenstern, and Herbert Shapiro. "If You Don't Know How Children Think, How Can You Help Them?" *Arithmetic Teacher* 22 (November 1975): 580–85.

Reisman, Fredricka K. *Diagnostic Teaching of Elementary Mathematics: Methods and Content.* Skokie, Ill.: Rand McNally & Co., 1977.

_____. *A Guide to the Diagnostic Teaching of Arithmetic.* Columbus, Ohio: Charles E. Merrill Publishing Co., 1972.

12

Estimation and Mental Arithmetic: Important Components of Computation

Paul R. Trafton

PUT yourself in each of the following situations, typical of those encountered daily.

300 × 4 = 1200, so that's a ratio of 4 to 1; 250 × 5 = 1250, for a ratio of 5 to 1. The ratio of advancing to declining stocks is between 4 to 1 and 5 to 1.

1206 stocks advanced on the market and 275 declined.

196

To roughly determine your gas mileage, you could think:

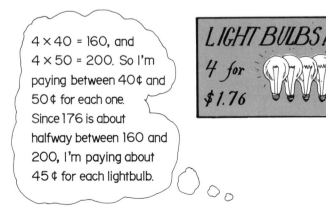

Miles driven	Gallons	Cost	M.P.G.
248	12.8	$9.37	

Actually this estimate could be adjusted by computing 12.8 × 20 = 256 miles. A 20 miles a gallon average was missed by 8 miles. Further refinement could lead to an estimate of just under 19.5 miles per gallon.

To find how much you are paying for each light bulb, think:

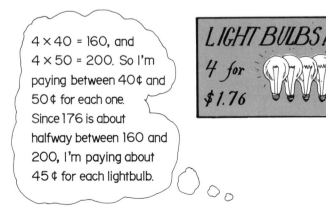

A ball player has 5 hits in 17 times at bat. You think:

$\frac{5}{15} = \frac{1}{3}$, so he is batting less than .333 — or, $\frac{1}{10}$ of 17
is 1.7, so $\frac{3}{10}$ of 17 is 5.1. His average is just about .300.
I wonder what his average will be if he goes
2 for 3 today?

It is now 2:15 P.M. Is it possible to be in Boise for a 5:00 P.M. meeting, assuming I average 55 mph?

Two 50s is 100, so 2 × 55 = 110.
In 2 hours I can go 110 miles; that leaves
25 miles, which can be traveled in . . . about
30 minutes.
    Let's see — 2:15 . . . 4:15, and 30 minutes
more is 4:45. Whew, I can just make it!

BOISE
135 mi

Suppose two tapes were purchased along with a pen. The cash register whirred and clicked, automatically adding in the sales tax, and "announced" a total of $4.79.

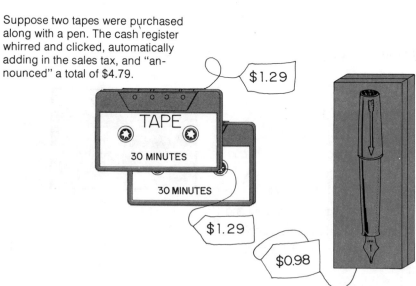

$1.29

TAPE

30 MINUTES

30 MINUTES

$1.29

$0.98

By several approaches you should be able to determine that something is wrong. Some possible thought patterns follow:

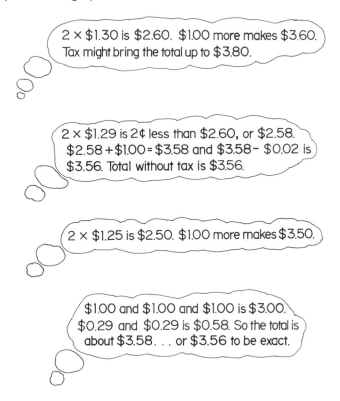

Such situations call for the use of estimation skills or the mental computing of exact answers through nonstandard techniques. Because of the usefulness of these techniques, it is most important that proficiency in estimation and mental computation be included as goals for the study of computation. Although the importance of estimation and mental arithmetic has long been recognized, these topics are frequently ignored in curriculum materials and classrooms or treated in a superficial, stilted, isolated manner. Despite the fact that estimation is often most useful in applied situations (where it is frequently accepted in lieu of an exact answer), many times it is studied only as a tool for checking the reasonableness of written calculation.

Today there is a new surge of interest in estimation and mental arithmetic. The case for proficiency with these skills is a practical one; they are widely used in many situations and thus form part of the minimal competence in mathematics expected of all citizens. Estimation and mental arithmetic contribute to the mathematics curriculum in additional ways. First, they can bring a new dimension and vitality to the study of computation. This is particularly true in upper grades, where students review familiar skills and focus on more complex levels of computation. Second, the work probably contributes to the development of quantitative thinking. Students become more adept at reasoning with numbers and more flexible in their thinking with numbers, gain new insights into operations and number relationships, and possibly develop a greater feeling of command over numbers. Third, the work develops problem-solving skills. For a given situation students must make decisions about how to treat the numbers, select a computational strategy that fits the situation, and employ a series of steps for the approach selected.

It is important to note that despite the conjectured values to be gained from this aspect of computation, little is known about the development of students' thinking in this area or about ways of helping them become competent and confident with these skills. However, it is clear that some approaches are more effective than others, that students do grow in their ability when systematic instruction is provided, and that many teachers are successful in making this kind of learning a satisfying, productive experience.

Strategies for developing estimating skills are presented in the next section. Then approaches to mental arithmetic are discussed. Finally, guidelines are given for effective development of the work in the classroom.

Developing Estimation Skills

Estimation is often viewed as a unitary skill. A broader perspective is useful, in which a variety of approaches and contexts for estimation are accepted. Although the four approaches treated in this section are somewhat interrelated, they do differ, and each tends to fit certain types of situations particularly well.

Computing with rounded numbers

	Example		Estimate
This is the most familiar approach, and it is frequently the only one developed. The appropriate numbers are rounded to the nearest ten, hundred, thousand, and so forth, and the rounded numbers are used in the computation.	362 + 825	⇨	400 + 800
	49 × 6	⇨	6 × 50
	68 × 32	⇨	70 × 30
	1782 ÷ 3	⇨	1800 ÷ 3

The ability to round numbers, a place-value skill, is basic to this type of estimation. In fact, a strong understanding of place value is important to all work with estimation and mental arithmetic. The idea of rounding numbers needs careful development and frequent review before students can apply it effectively to work in estimating answers. Before learning the usual mechanical shortcuts for rounding numbers, they need to understand the reasoning pattern and the steps employed. Key steps for rounding 268 to the nearest hundred follow. They are illustrated in figure 12.1.

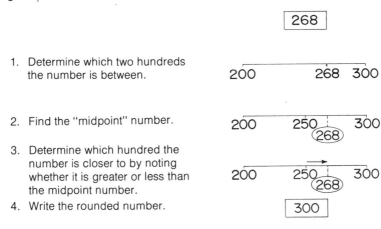

1. Determine which two hundreds the number is between.

2. Find the "midpoint" number.

3. Determine which hundred the number is closer to by noting whether it is greater or less than the midpoint number.

4. Write the rounded number.

Fig. 12.1

Students should see that this process applies to all rounding situations. Also, they need to learn at least one of the rules for dealing with midpoint numbers.

Many children find this approach the most difficult to use and to develop confidence in, particularly when they are asked to use it along with computing exact answers. They may find conventional computation in addition and subtraction easier than rounding. It is often easier for teachers to get them to use rounding for multiplication, since estimation in multiplication is easier than the usual computation.

Three suggestions are provided for making the work more effective:

1. Have students find estimates only. This helps them focus just on the estimation task. It also fits the way estimation is often used in daily life.

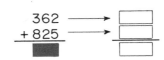

2. Provide exercises where students select the example that gives the better estimate.

$5 \times 67 = \square$ a. 5×60 b. 5×70

$42 \times 48 = \square$ a. 40×40 b. 40×50

3. When combining estimation with computing exact answers, have students first make all estimates under your direction before doing the conventional computation.

Using a reference point

Can you buy both with $1.00?

Can you buy 4 for $5.00? What's the most you can buy with $5.00?

Is the difference in cost more than $1.00?

Am I paying more than $1.00 for each ball or less than $1.00?

These four situations incorporate the use of a fixed amount into the estimation process. The goal is to determine whether the computation will be "over" or "under" the reference point. Many times the estimate is not a number, as shown in figure 12.2. Here the process ends as soon as it is determined that the sum is greater than $5.00. The techniques the estimator uses in the reasoning process may involve computing with rounded numbers. However, the situation itself and the presence of a reference number often promote a variety of reasoning patterns.

"Can I buy both for $5.00?"

Fig. 12.2

Thus, estimation using a reference point tends to change the reasoning, so that techniques other than computing with rounded numbers are widely used. The presence of the reference point provides additional guidance and can make the work easier. The four illustrations at the beginning of this section suggest that this type of estimation is particularly useful in situations involving money.

The use of a reference point suggests exercises that promote reflective thinking about computation and build a base for more formal work in estimation. In the exercises that follow, the student is given an estimate and asked to compare other exercises to it.

1. $\underline{50 + 20 = 70}$; so 52 + 23 is _____ than 70. (greater, less)
   $\underline{80 - 30 = 50}$; so 80 − 36 is _____ than 50. (greater, less)

2. $\underline{400 + 400 = 800}$ Which are greater than 800? Which are less?

408 + 419	400 + 375
389 + 397	400 + 415

3. $\underline{6 \times 70 = 420}$ Which are greater than 420? Which are less?

6 × 72	6 × 69
6 × 67	6 × 73

4. $\underline{4200 \div 7 = 600}$ Is 4230 ÷ 7 greater or less than 600?

5. $\underline{1/2 + 1/2 = 1}$ Which are greater than 1? Which are less?

$$1/2 + 3/8 \qquad\qquad 1/2 + 5/8$$

$$3/8 + 3/8 \qquad\qquad 5/8 + 4/8$$

"Front end" estimation

A particularly appropriate approach to estimating answers in addition and subtraction with numbers of three or more digits is illustrated in figure 12.3.

426		
275	*a.* Add the hundreds ⟶	$400 + 200 + 100 = 700$
+126	*b.* Estimate the sum of ⟶ tens and ones	greater than 100 but less than 200
	c. Estimate the total ⟶	between 800 and 900

626	*a.* Subtract the hundreds ⟶	$600 - 100 = 500$
−147	*b.* Compare 26 and 47 ⟶	26 is less than 47
	c. Estimate the difference ⟶	The answer is between 400 and 500.

Fig. 12.3

In this approach the numbers are not rounded. Rather, the leftmost digits are added or subtracted. Then by "eyeballing" the remaining digits, one can obtain an estimate for this part of the problem. Finally, an estimate for the example is determined. The example

$$\begin{array}{r} \$6.95 \\ 5.14 \\ +2.79 \\ \end{array}$$

is solved first by rounding and then by front-end addition:

Rounding	Front–End Addition
$7.00	$6.00 + $5.00 + $2.00 = $13.00
5.00	$0.95 and $0.14 is over $1.00 and
+ 3.00	$0.79 makes it close to $2.00. So
$15.00	the estimate is $15.00.

Both approaches give the same estimate in this example, and both are appropriate techniques. One possible advantage of the front-end technique is that it is less likely to require the use of pencil and paper than the rounding approach. A value of being proficient with both approaches is that one can choose the technique that fits a given situation best.

The front-end approach can also be employed for some multiplication examples.

$$4 \times 648 = \square$$

a. $4 \times 600 = 2400$
b. 48 is about 50
c. $4 \times 50 = 200$
d. the product is about 2600

It is interesting to note that our traditional approach to long division is, in effect, a front-end approach, since we find the digit with the greatest place value in the quotient first.

Estimating the range of the answer

- A field measures 200 meters by 100 meters. Is its area 2 000 m², 20 000 m², or 200 000 m²?
- A pilot announces that the plane's altitude is 36 000 feet. Since 36 divided by 5 is about 7, is the plane's altitude about 7 miles or 70 miles?
- A house is selling for $70 000. The realtor's commission is 6%. Will the realtor's share be $420, $4 200, or $42 000?
- You multiply 36 by 69.2 on a calculator. It displays 2491.2 as the product. Is this correct, or should it be 249.12 or 24.912?

In each of these situations the appropriate size of the answer or correct order of magnitude must be identified. In other estimation work the goal is to determine an answer that is reasonable in the sense of being close to the actual answer. Here the goal is to find if an answer is reasonable in the sense of being in the right range. It involves finding whether the answer should be in the ones, in the tens, or in the hundreds, for example, or if the decimal point has been positioned correctly. It is not uncommon for students and adults to state that 320 is the product of 40 and 80. Knowing that the product of two multiples of ten must be a multiple of 100 would prevent this error. It is also true that the product of two two-place numbers will be either a three-place or four-place number.

Determining the correct order of magnitude when large numbers are involved is the result of computing with the significant digits, which represents a form of front-end computation. Place-value understanding and skill in multiplying and dividing by powers of ten are important components.

This type of estimation skill takes on added importance with the widespread use of the hand-held calculator. It is easy to press a wrong key or an extra key or fail to enter a zero or a decimal point. Since many calculators do not produce printouts, skill in estimating an answer's order of magnitude, as well as in estimating reasonable answers within the right order of magnitude, takes on increased importance.

This type of estimation is important when a series of computations is involved. In each example that follows, it is possible to estimate an answer close to the actual one. Yet perhaps only the answer's correct order of magnitude is crucial, particularly if one is using a calculator. At the least, a person should sense that something is wrong with a drastically incorrect answer.

$$\frac{(36 \times 24) + 368}{16} \longrightarrow \frac{(40 \times 20) + 400}{16} \longrightarrow \frac{1200}{16}$$

At this point it is clear that a one-place or four-place answer is inappropriate. But one can go further: "There are at least ten sixteens in 1200 but not as many as a hundred sixteens. The answer is in the tens."

$$\frac{(47.3 + 6.25 + 260) \times 11}{15.6} \longrightarrow \frac{320 \times 11}{16}$$

Here the estimator might think

$$\frac{3200}{16}$$

or reason; "16 'goes into' 320 twenty times; so 20 × 11 = 220."

Finally, this type of estimation is important in multiplying and dividing with decimals as well as in doing long division. In the example

$$\begin{array}{r} 4.6 \\ \times\ 3.5 \\ \hline 1610 \end{array}$$

students should be able to place the decimal point correctly without relying on the usual mechanical rule. That the answer is "in the teens" should be obvious.

In computing a long division example, knowing the number of places in the quotient is an important part of finding a reasonable answer and avoiding common errors in using the algorithm. An illustration follows:

34)2740   10 × 34 = 340 There are at least 10 thirty-fours in 2740.
            100 × 34 = 3400 There are not as many as 100 thirty-fours in 2740.

So the quotient is between 10 and 100. There will be two digits in the quotient.

This technique is particularly useful when zeros occur in the quotient:

$$\begin{array}{r} 7\ 0\ 9 \\ 6\overline{)4\ 2\ 5\ 4} \end{array}$$

Developing Mental Arithmetic Skills

The term *mental arithmetic* is often used to refer to nonstandard algorithms for computing exact answers. The computation is done without pencil and paper. In a sense the term is a misnomer, since to some extent all arithmetic is done mentally. Nonetheless, the term has wide acceptance.

Many procedures of mental arithmetic have been devised, including several elaborate ones for multiplying larger numbers. The work in this section is limited to simple addition and subtraction situations together with one type of multiplication example. This level of mental arithmetic is the most useful in daily life and is often used in conjunction with an estimation procedure (as illustrated by the final example in the introduction to the article).

Four alternatives to the standard algorithm for addition and one for subtraction follow. These algorithms are as systematic as the conventional ones; yet their front-end nature permits easier remembering of partial answers and more meaningful interpretation of the numbers themselves. Although many adults have become adept at using the conventional algorithms without paper and pencil, these approaches tend to be more efficient for mental computation. They are also more appealing to many students than just estimating with two-place numbers.

④8 + ③5

1. Add the tens.	40 and 30 is 70
2. Add the ones.	8 and 5 is 13
3. Add the two sums.	70 + 13 = 83

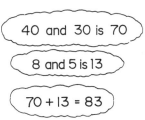

(48) + ③5

1. Look at the whole first number. Add to it the tens in the second number.	48 and 30 is 78
2. Add the ones in the second number to the sum.	78 + 5 = 83

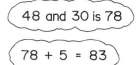

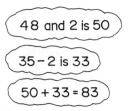

1. Add a number to one addend to make it a multiple of ten.

48 and 2 is 50

2. Subtract the same number from the other addend.

35 − 2 is 33

3. Add the two numbers.

50 + 33 = 83

(48) + 19   (for use in adding 19, 29, etc.)

1. To the first number add 20 (the next higher multiple of ten).

48 and 20 is 68

2. Subtract 1 from the sum.

68 − 1 = 67

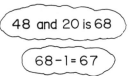

1. From the first number subtract the tens in the second number.

62 minus 20 is 42

2. Now subtract the ones in the second number.

42 − 5 = 37

Two situations for multiplication are now presented. The first represents a front-end approach to the conventional algorithm. The second implicitly involves the use of the distributive property of multiplication over subtraction. Both can be employed as a means of moving beyond an initial estimate to the exact answer.

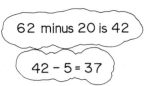

6 × 72

1. Multiply the tens in the second factor by the first factor.

6 × 70 = 420

2. Multiply the ones in the second factor by the first factor.

6 × 2 = 12

3. Add the two products.

6 × 72 = 432

$\boxed{6 \times 39}$

1. Round the second factor to the next higher ten.

$39 \longrightarrow 40$

2. Multiply the tens by the first factor.

$6 \times 40 = 240$

3. Multiply the difference by the first factor.

$40 - 39 = 1$

$6 \times 1 = 6$

4. Subtract the second product from the first one.

$240 - 6 = 234$

Getting started

Two approaches for beginning work with mental computation follow. Both can be used in primary grades as well as with older students.

1. When you introduce the addition of multiples of ten, present worksheets showing the following equations and drawings. Have models available also.

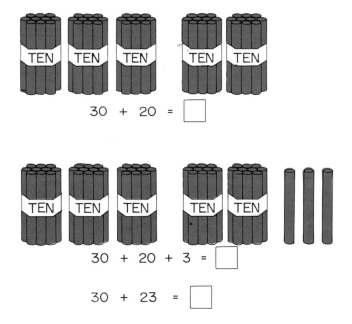

$30 + 20 = \boxed{}$

$30 + 20 + 3 = \boxed{}$

$30 + 23 = \boxed{}$

After the initial presentation, the work can be shortened and the use of models becomes optional.

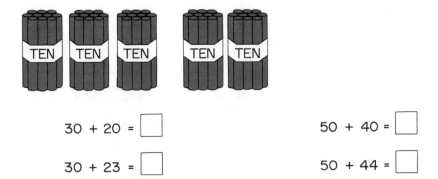

30 + 20 = ☐

30 + 23 = ☐

50 + 40 = ☐

50 + 44 = ☐

2. Have students use a hundreds chart to find numbers that are ten greater than a given number. They should see, for example, that the number that is 10 greater than 15 is one row below it (see fig. 12.4). Then help them see that finding the number that is 10 greater than 15 is equivalent to finding the sum of 15 and 10. Give several practice exercises in which students *(a)* locate a number and *(b)* find the sum of that number and 10. *(Example:* Find 28. Now show the sum for 28 + 10.) When they are proficient at this, move to finding the number that is 10 less than a given number. Repeat the approach, this time adding 20 and 30 to a number as well as subtracting 20 and 30 from the number.

1	2	3	4	5	6	7	8	9	10
11	12	13	14	15	16	17	18	19	20
21	22	23	24	25	26	27	28	29	30
31	32	33	34	35	36	37	38	39	40

Fig. 12.4

One additional step is to have students solve examples such as 15 + 23 using the procedure shown in figure 12.5. Both of these activities encourage students to add a number of tens to the whole first addend.

1	2	3	4	5	6	7	8	9	10
11	12	13	14	15	16	17	18	19	20
21	22	23	24	25	26	27	28	29	30
31	32	33	34	35	36	37	38	39	40

You have taken a brief look at nonstandard approaches for certain types of addition, subtraction, and multiplication examples. The work differs markedly from standard approaches. Sometimes one number is viewed as a whole instead of dealing with tens and ones places, and in other instances tens are used in computing before ones. Each of the techniques facilitates "grocery store" computation, and often the technique permits moving beyond an initial estimate to an exact answer. From these elementary stages more advanced work can be developed with larger numbers, additional strategies can be devised, and older children can profit from discussing the reasoning underlying each approach.

Guidelines for Classroom Instruction

In previous sections, types of estimation and mental arithmetic as well as selected strategies have been presented. In this section guidelines for developing this aspect of computation are given. Since little curriculum material exists that treats this work in a consistent, comprehensive manner, responsibility is placed on teachers for instructional planning and implementation.

Make procedures and skills in estimation and mental arithmetic an objective for all. A strong case exists for making this work part of the basic mathematical competence for all students and adults. It is widely used in a variety of ways and is particularly important in on-the-spot computation by consumers.

Recognize that it needs to be taught. The majority of students need planned, systematic instruction in order to develop both the necessary skills and confidence in using them.

Build basic number ideas, concepts of operations, and computational skills meaningfully. Techniques of estimation and mental arithmetic involve high-level ways of interpreting numbers, manipulating numbers, and reasoning using alternative ways of computing. These procedures are difficult to develop if pupils' other work with numbers has been limited to mechanical procedures that have little meaning. Individuals for whom fractions and decimals, for example, have little meaning will have great difficulty in moving to more complex avenues of thought. Similarly, if an operation, such as division, has little meaning, students will have difficulty in thinking flexibly about division situations.

Stress mathematical reasoning throughout computation. Confidence in reasoning with numbers builds over time. In early work the use of thinking strategies for basic facts not only facilitates the learning of facts but also builds a foundation for other reasoning strategies:

$6 + 7 = \square$ $8 \times 6 = \square$

6 and 6 is 12; so 6 and 7 is 1   I know 4 sixes is 24; so 8 sixes is
more: $6 + 7 = 13$ twice as much: $8 \times 6 = 48$.

Primary-grade children can easily extend these patterns to examples like
"$30 + 30 = 60$; so $30 + 31 = ?$" and "$25 + 25 = 50$; so $25 + 26 = ?$."
Exercises like the following promote a reflective approach to computation
and fit well into later estimation work:

1. Which is greater, 10×35 or 360?
2. $8 \times 15 = 120$; so $16 \times 15 = ?$
3. $3 \times 24 = 72$; so $30 \times 24 = ?$
4. $6 \times 7 = 42$; so 6×7 tens $= ?$ tens; $6 \times 70 = \square$

Make the work an integral part of the instructional program. Students
make the greatest growth in estimation and mental arithmetic when these
skills are taught as a regular, ongoing part of the mathematics program.
This involves planning lessons that focus on the development of specific
strategies and thinking patterns. Regular emphasis is of equal importance.
As students have regular contact with this work, they more easily accept
it as a normal part of computation, have repeated opportunity to become
more proficient, and are better able to integrate it with the conventional
aspects of computation instruction, which involve quite different ways of
thinking.

*Place estimation and mental arithmetic in the context of application
situations.* The two broad areas in which this work is of value are (1) as
an internalized check on written or calculator computation and (2) in
social and other situations in which estimation and mental arithmetic replace or at least precede written computation. It has been suggested that
a large portion of our daily use of numbers involves estimation rather than
actual computation. Thus, putting the instructional work into different
contexts makes the school work fit closely the situations in which students will apply it. Application settings also provide a strong motivation
for the work as its relevance is clearly established. Throughout this article
many of the examples have been drawn from common, everyday experiences. Such situations abound and can easily be drawn on for instruction if teachers develop an awareness of them.

Stress oral work. Although paper-and-pencil experiences can be useful, class discussion is a more powerful way of developing ideas. Many
students have greater confidence and skill in written computation and will
resort to this method initially if the situation permits, basing their estimate
on the exact answer. Initial resistance to estimation and mental arithmetic

is to be expected. For one thing, it differs greatly from working with conventional algorithms, which consist of breaking examples into small pieces. (Many students see a series of small steps and ignore the total example.) Using many oral situations regularly reduces their anxiety about errors, promotes greater freedom to experiment, and allows them to work out their thinking aloud, as well as to listen to the thinking patterns of others.

Accept individual differences. Students differ greatly in their ability to use these techniques, in the level of thinking employed, and in the procedures by which they arrive at solutions. This is to be expected. Accept immature or stumbling efforts even as you encourage other levels of performance. The development of proficiency is a long-range goal of instruction. With respect to procedures employed, encourage different approaches and accept different solutions. Some students will compute with rounded numbers only. Others will attempt to adjust a rough estimate to make it more precise. A few will create elaborate techniques representing a blend of estimation and mental arithmetic procedures that result in excellent estimates or even exact answers. It is this diversity that makes this work so interesting to do with students.

Summary

This article has presented selected aspects of the topics of estimation and mental arithmetic, examined their role in the study of computation, and provided a discussion of procedures for developing proficiency and confidence in this area. The purpose has been to provide direct, practical help to those working directly with students. Since this topic is a broad one, only selected aspects have been presented, and the discussion has been brief. The work with estimation has been limited to its use in computation, although it also has an important role in the study of measurement. Many techniques of mental arithmetic exist beyond those that have been presented here, and some aspects of estimation in computation—namely, its use in carrying out long division—have not been included.

It is to be hoped that more will be learned in the next few years about how students develop these skills, how this work can best be integrated into the curriculum, and how instruction can more closely fit the psychology of the learner. Nonetheless, the significance of the work is clear, and evidence suggests that approaches built around the ideas presented in this article are effective. It is important that we proceed in light of the knowledge we possess. Certainly teachers and students alike will profit from this work and enjoy making it an integral part of computation at all levels.

13

Computation and More

Diane J. Thomas

WHAT is new or exciting about computation? For many junior high school students the answer too often is, "Nothing." A been-here-before attitude on their part, perhaps coupled with memories of a lack of success, is at the heart of the teaching problem in developing computational skills at the secondary school level. Certainly students need clear explanations of computational procedures and plenty of opportunity for drill and practice. But the approach that a teacher uses is all-important in establishing a positive framework for these computation activities.

One way to catch the interest of students is to introduce them to new, but easy, ideas from different areas of mathematics so that some of the needed computation review and practice can be embedded within investigations of higher-level concepts. Along with being exposed to fresh mathematical ideas, students sharpen their computational skills in a two-birds-with-one-stone approach.

Following are some examples of topics and activities accessible to junior high school students and to general mathematics students in high school. In each of these topics—density of the rational numbers, perfect squares and square roots, and modular arithmetic—computation acts as a needed tool within a larger problem-solving context. Much of the motivation for the activities comes from the problem-solving situation itself—students of all ages and ability levels enjoy puzzling over questions that do not seem too forbidding, especially if the lead-up questions have been easy to answer. Looking for patterns and guessing what will happen next are other activities that students usually enjoy. Each of the three topics has been organized to capitalize on student curiosity about numbers and number patterns.

Density of the Rational Numbers

What is the number halfway between 7 and 15? What about the number halfway between 28 and 191? What if two given numbers are fractions—say, 1/4 and 2/3—will there be a number halfway between? What number? How can we find it?

The density property of the rational numbers can be used as an organizer for helping students review fractions and practice computational skills. Starting out witti easy problems involving whole numbers, students often quickly find the rational number halfway between two given whole numbers by counting or by referring to a number line: 11 must be the number halfway between 7 and 15, since $7 + 4 = 11$ and $11 + 4 = 15$. The problem is represented on the number line in figure 13.1. This process foreshadows the definition of *midpoint*, which students will encounter later in geometry: C is the midpoint of $\overline{AB}$ if and only if C is between A and B, and $AC = CB$.

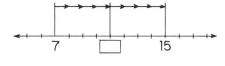

Fig. 13.1

The counting technique or the number-line picture is an easy method for locating the midpoint when the distance between the two given numbers is small. For the problem of finding the number halfway between 28 and 191, however, a different method is needed to locate it efficiently. Students generally discover that one procedure is to find the distance (or difference) between 28 and 191:

$191 - 28 = 163$

Take half that distance, since the midpoint is desired:

$163 \div 2 = 81\frac{1}{2}$

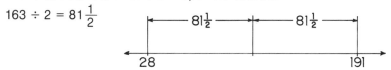

Add that value to the smaller number:

$$28 + 81\frac{1}{2} = 109\frac{1}{2}$$

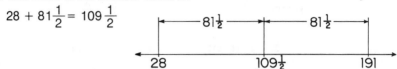

(Notice that subtraction is viewed as finding the directed distance between two numbers, a concept used in later mathematics courses with problems such as $|x - 3| = 8$.) Another procedure, not quite as obvious to students, is averaging the two given numbers to find the number halfway between them:

$$\frac{28 + 191}{2} = \frac{219}{2} = 109\frac{1}{2}$$

The same approaches can be used when the two given numbers are fractions. First, easy problems that can be solved by counting on a number line are presented:

1. Find the number halfway between 1/8 and 5/8.

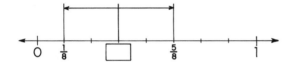

Answer: 3/8

2. Find the number halfway between 3/20 and 13/20.

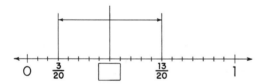

Answer: 8/20, or 2/5

3. Find the number halfway between 1/2 and 7/10.

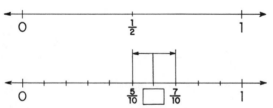

Answer: 6/10, or 3/5

In the process of solving these problems, students review how to locate fractions on the number line *and* how to work with equivalent fractions using a number-line model.

Next, problems like finding the number halfway between 3/5 and 4/5 can be attacked using the number line:

Students almost always say "$3\frac{1}{2}$-fifths" — the idea of a complex fraction arises naturally. When the number-line picture is redrawn to include the midpoints of each of the original intervals, it becomes evident that $3\frac{1}{2}$-fifths can be renamed as the more "usual" fraction, 7/10:

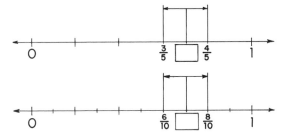

The geometric idea involved is that of refining a partition, a concept used in integral calculus as well as in more prosaic topics such as linear measurement (greatest possible error). Students are usually comfortable with the idea of multiplying a common fraction by different forms of the number 1 to obtain an equivalent fraction:

$$\frac{3}{5} = \frac{3}{5} \times 1 = \frac{3}{5} \times \frac{2}{2} = \frac{6}{10}$$

(In this example, the multiplication by 2/2 may be connected with the geometric process of subdividing each interval into two halves.) There does not seem to be much difficulty in extending the multiplication procedure for use with complex fractions as well:

$$\frac{3\frac{1}{2}}{5} = \frac{3\frac{1}{2}}{5} \times 1 = \frac{3\frac{1}{2}}{5} \times \frac{2}{2} = \frac{7}{10}$$

Also, a facility with complex fractions is useful later when the algorithm for dividing fractions is developed. Rewriting the problem $1/3 \div 7/8$ as the complex fraction

$$\frac{\frac{1}{3}}{\frac{7}{8}}$$

and then multiplying both terms of the fraction by

$$\frac{\frac{8}{7}}{\frac{8}{7}}$$

may be confusing to students if they have not previously encountered complex fractions or the multiplication procedure for simplifying complex fractions.

The general procedures for finding the number halfway between two whole numbers can be applied to problems concerning fractions. For example, to find the number halfway between 1/4 and 2/3, the following processes can be used:

1. First, find the distance between 1/4 and 2/3 by subtracting:

    2/3 − 1/4 = 8/12 − 3/12 = 5/12

    Then take half that distance, since the midpoint is desired:

    1/2 of 5/12 = 1/2 × 5/12 = 5/24

    Add the result to the smaller fraction:

    1/4 + 5/24 = 6/24 + 5/24 = 11/24

    Thus, the number halfway between 1/4 and 2/3 is 11/24.

2. Or use the averaging method:

$$(1/4 + 2/3) \div 2 = (3/12 + 8/12) \div 2$$
$$= 11/12 \div 2$$
$$= 11/12 \times 1/2$$
$$= 11/24$$

3. Or since 1/4 = 3/12 and 2/3 = 8/12, the number halfway between 3/12 and 8/12 must be $\frac{5\frac{1}{2}}{12}$, which is equivalent to

$$\frac{5\frac{1}{2}}{12} \times \frac{2}{2} = \frac{11}{24}.$$

Computational skills for addition and subtraction get reviewed when the first two methods are used; in all three methods multiplication and division of fractions are reintroduced gently, since only 1/2s and 2s are involved as multipliers or divisors. Further, a discussion of multiple approaches to solving the same problem is beneficial for students.

Is there always a whole number halfway between any two whole numbers? Can one always find a number halfway between any two fractions? Will this middle number always be a rational number? How many rational numbers are there in a given interval on the number line? Questions like these, accompanied by appropriate computation problems, can get students thinking about the density property for the rational numbers.

Perfect Squares and Square Roots

Many students seem to be fascinated by the idea of perfect squares and square roots. The advent of the hand-held calculator, with its nearly instantaneous square root key, does not negate the need to teach a procedure in which the *concept* of square root is emphasized. When the buildup to the topic of square roots is gradual and when the memorization of complex algorithms is avoided, even general mathematics students respond positively to this topic.

A review of the basic multiplication facts that lead to perfect squares is a good beginning. Then, once students remember facts like $7 \times 7 = 49$ and $9 \times 9 = 81$ and once they also know that $70 \times 70 = 4900$ and $90 \times 90 = 8100$, they can start talking about the inverse operation of taking square roots. (If this topic is approached early in the school year when whole-number computation is being developed, it is the principal square root with which students will be dealing; later, after they have studied integers, perhaps a return to the square root problem would be appropriate.) Simple problems like $\sqrt{36} = 6$ and $\sqrt{6400} = 80$ can lead to more difficult problems, such as finding the square root of the perfect square 1849. Here, teaching students a method of attack is useful. If there is a list such as the one in figure 13.2 already on the blackboard,

Square Root	Perfect Square
1	1
2	4
3	9
.	.
.	.
.	.
10	100
20	400
30	900
40	1600
50	2500
60	3600
70	4900
80	6400
90	8100

Fig. 13.2

students can be led to suggest that 1849 will fit between 1600 and 2500 in the "perfect square" column:

Square Root	Perfect Square
40	1600
	←1849
50	2500

Thus a likely guess for its square root would be some number between 40 and 50. At first, a guess-and-check procedure can be used. (Multiplying the guessed answer by itself to see if the product really is 1849 seems to be preferred by students over checking by long division.) Later, procedures such as the following that narrow down the possible square root choices can be discussed:

1. *Analyze the last digit of the given perfect square.* For example, if the perfect square is 1849, its units digit is 9. Only numbers ending in 3 or in 7 will yield a units digit of 9 when squared. Thus, since the guess for the square root must be some number between 40 and 50, the numbers 43 and 47 are the candidates.

2. *Look at the position of the perfect square with relation to the known perfect squares.* Since 1849 is closer to 1600 than to 2500, it is reasonable to guess that the square root of 1849 is closer to the square root of 1600 (i.e., 40) than to the square root of 2500 (i.e., 50). Therefore the choice of 43 seems to be sensible.

A series of exercises in finding the principal square root of one-, two-, three-, and four-digit perfect squares will give students practice in multiplication (or in long division if that method of checking is preferred). They also get to apply the ideas of "greater than," "less than," and "betweenness" as an aid to estimation. Logical reasoning and analysis skills come into play as students become concerned with narrowing down their number of guesses (and thus reducing the amount of work in checking each guess).

After working a few of these square-root-of-perfect-squares problems, someone in the class always will be bothered that the numbers come out so well and ask, "What happens when you run into a number that doesn't work?" What a beautiful question! The concept of irrational number can now be introduced naturally in a guided-discovery manner:

1. Have the questioner select a fairly small number that probably "won't work," for example, 19.
2. Use the method of attack described earlier.

   a) Locate the given number between two of the perfect squares:

Square Root	Perfect Square
4	16
	←19
5	25

Here it appears that some number between 4 and 5 would be a likely candidate for the square root of 19.

b) Pick some number within the determined range, and check it:

In this example, students will probably suggest $4\frac{1}{2}$. Checking $4\frac{1}{2}$ as a possible square root yields

$4\frac{1}{2} \times 4\frac{1}{2} = 9/2 \times 9/2 = 81/4 = 20\frac{1}{4}$.

Since $20\frac{1}{4}$ is larger than the sought-after 19, the guess must have been too large. The guessing range is thus reduced to numbers between 4 and $4\frac{1}{2}$.

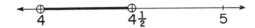

c) Repeat the process as long as the students have patience.

Maybe $4\frac{1}{4}$? $4\frac{1}{4} \times 4\frac{1}{4} = 17/4 \times 17/4 = 289/16 = 18\frac{1}{16}$, which is smaller than 19 Now we know that the square root of 19 must lie somewhere between the numbers $4\frac{1}{4}$ and $4\frac{1}{2}$.

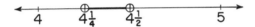

Wouldn't using decimals simplify all these calculations? Next guess?

3. Sometime the question "Will we ever get a number so that when we multiply it by itself we will get the answer 19?" needs to be raised and discussed.

In working these square root problems, students have a practical reason for needing to know how to multiply mixed numbers. If they are tempted to use a shortcut in step 2b above ("Since 4×4 is 16 and $1/2 \times 1/2$ is 1/4, then $4\frac{1}{2} \times 4\frac{1}{2}$ must be $16\frac{1}{4}$" is their usual line of reasoning), a refresher concerning the distributive law applied to $(4 + 1/2) \times (4 + 1/2)$ might be helpful in showing why that shortcut does not hold. In deciding which new number to test at each step, students review the concepts of order and betweenness of points on the number line. They must also deal with equivalent forms of a number: If the square root of 19 lies between $4\frac{1}{4}$ and $4\frac{1}{2}$, what is the next number that could be tested? There are several choices (fig. 13.3).

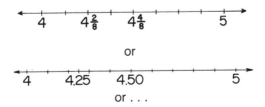

Fig. 13.3

The pros and cons of choosing the midpoint of an interval can make an interesting discussion. And throughout all the searching for square roots, students see that guessing—then checking and refining the guess—can be a legitimate problem-solving approach in mathematics.

Modular Arithmetic

"Mod arithmetic" is another area rich in computation practice for students. Even the name itself—especially the "mod" part—is tantalizing.

When the clock is used as a model ("What time shows on the clock when it is six hours after midnight? Thirteen hours? Twenty-four hours? Thirty-nine hours?"), students quickly learn how numbers are represented in a mod-12 system. From simple counting to basic addition and multiplication facts in mod-12 is an easy step for most students. Generalizing to a modular system different from 12 is not very difficult if "clocks" are used: a mod-7 system would mean that the clock would be numbered from 1 through 7, and arithmetic operations would be conducted within that restricted range of seven elements. (See Mueller [1963] for further details.)

Looking at the operation of multiplication in many different modular systems all at once leads to a good exercise in finding and predicting number patterns. For example, students can be asked to fill in the multiplication tables for mod-3 through mod-8. Figure 13.4 shows the completed tables.

In completing the tables, students obviously get practice in basic multiplication, division, and subtraction facts. Moreover, many of them will notice and use number patterns to lessen the amount of work involved in filling in the tables. Some of the common patterns among the tables that students usually find are these:

1. In each of the tables, a row matches the corresponding column. For example, in the mod-5 table, row 3 across reads 3-1-4-2-5, and column 3 down also reads 3-1-4-2-5. This is really just another way of saying that multiplication in a modular system is commutative.

2. In each of the tables, the element "1" acts as the multiplicative identity.

3. In the last row (and column) of each of the tables, the base of that modular system is repeated in each of the cells. For example, in the mod-4 table, row 4 across reads 4-4-4-4; in the mod-5 table, row 5 reads 5-5-5-5-5; and so on. The base of the modular system, then, acts just like zero does in our regular number system.

From here, students' attention can be directed toward the differences between some of the tables. Students readily point out that in some of the tables each element of that modular system appears once and only once in every row except the last row (for example, in the mod-7 table, each

Mod 3

×	1	2	3
1	1	2	3
2	2	1	3
3	3	3	3

Mod 4

×	1	2	3	4
1	1	2	3	4
2	2	4	2	4
3	3	2	1	4
4	4	4	4	4

Mod 5

×	1	2	3	4	5
1	1	2	3	4	5
2	2	4	1	3	5
3	3	1	4	2	5
4	4	3	2	1	5
5	5	5	5	5	5

Mod 6

×	1	2	3	4	5	6
1	1	2	3	4	5	6
2	2	4	6	2	4	6
3	3	6	3	6	3	6
4	4	2	6	4	2	6
5	5	4	3	2	1	6
6	6	6	6	6	6	6

Mod 7

×	1	2	3	4	5	6	7
1	1	2	3	4	5	6	7
2	2	4	6	1	3	5	7
3	3	6	2	5	1	4	7
4	4	1	5	2	6	3	7
5	5	3	1	6	4	2	7
6	6	5	4	3	2	1	7
7	7	7	7	7	7	7	7

Mod 8

×	1	2	3	4	5	6	7	8
1	1	2	3	4	5	6	7	8
2	2	4	6	8	2	4	6	8
3	3	6	1	4	7	2	5	8
4	4	8	4	8	4	8	4	8
5	5	2	7	4	1	6	3	8
6	6	4	2	8	6	4	2	8
7	7	6	5	4	3	2	1	8
8	8	8	8	8	8	8	8	8

Fig. 13.4. Multiplication tables for mod-3 through mod-8

row except the last has the elements 1-2-3-4-5-6-7 arranged in some order), whereas in other tables, repeating patterns of elements appear within some of the rows (for example, in the mod-6 table, row 2 across reads 2-4-6-2-4-6). If the multiplication tables are classified as "once and only once" versus "some repeaters," an overall pattern may emerge:

"Once and only once"	"Some repeaters"
mod-3	mod-4
mod-5	mod-6
mod-7	mod-8

Students usually suggest that all tables with an odd number as the base will have once-and-only-once rows. This conjecture can easily be checked by completing a mod-9 table, as shown in figure 13.5. Since elements in rows 3 and 6 of the mod-9 table repeat in a pattern, the mod-9 table does not belong in the once-and-only-once category, even though 9 is an odd number. The next conjecture turns out to be correct: All tables with a *prime number* as the base will have once-and-only-once rows.

Examining the tables that have repeating patterns in some of their rows leads to other questions: Is there something special about the rows that have repeating patterns? That is, if someone picks a nonprime base for a mod system, say, mod-21, can we tell *without* completing a mod-21

Mod 9

×	1	2	3	4	5	6	7	8	9
1	1	2	3	4	5	6	7	8	9
2	2	4	6	8	1	3	5	7	9
3	3	6	9	3	6	9	3	6	9
4	4	8	3	7	2	6	1	5	9
5	5	1	6	2	7	3	8	4	9
6	6	3	9	6	3	9	6	3	9
7	7	5	3	1	8	6	4	2	9
8	8	7	6	5	4	3	2	1	9
9	9	9	9	9	9	9	9	9	9

Fig. 13.5. Multiplication table, mod-9

multiplication table which rows will have repeating patterns? At this stage, the data collected so far can be organized into a form that might make potential relationships more obvious:

Mod Table	Mod-4	Mod-6	Mod-8	Mod-9	Mod-21
Row number having a repeating pattern	2	2, 3, 4	2, 4, 6	3, 6	?

The first conjecture students usually make involves a "divides" relationship between the row number and the base of the modular system. The row number 2 divides the base 4; the row number 2 divides the base 6, the row number 3 divides the base 6—but the row number 4 does not divide the base 6. However, the original conjecture can be slightly modified to a relationship of the row number and the base sharing a common factor, and the overall pattern will hold. (Or students may prefer to think of a reduce-the-fraction situation, where the denominator of the fraction is the base of the mod system and the numerator of the fraction is the row number: 2/4; 2/6, 3/6, 4/6; 2/8, 4/8, 6/8; and so on. All these fractions are reducible because the numerator and denominator share a common factor.)

In this mod arithmetic unit, computation practice with basic facts and a review of properties such as prime number and common factor are done in a problem-solving context. Other important mathematical skills—looking for patterns, making and checking conjectures, using counterexamples—are exercised throughout the unit as well.

Summary

What is new or exciting about computation? A jaded "Nothing" need not be the students' response. There are occasions in which computation may be approached as an integral part of higher-level mathematical inquiries. The review and practice of computation make sense when the skills are considered a useful means of proceeding through some interesting new ideas. Naturally this approach does not remove the need for a systematic focus on computational procedures and their rationale. However, it can provide a fresh setting for reintroducing computational procedures or for giving students the needed maintenance activities. Set in a nonroutine context, the routine of computation is enlivened and enlarged.

BIBLIOGRAPHY

Listed below are several references, each suggesting other mathematical topics that might be used to set a context for computational activities.

Dunn, J. A. "Arrow Chains." *Mathematics Teaching* 62 (March 1973): 24–25.

McIntosh, Alistair, and Douglas Quadling. "Arithmogons." *Mathematics Teaching* 70 (March 1975): 18–23.

Maletsky, Evan M. "Ancient Babylonian Mathematics." *Mathematics Teacher* 69 (April 1976): 295–98.

Mueller, Francis J. "Modular Arithmetic." In *Enrichment Mathematics for the Grades,* pp. 73–91. Twenty-seventh Yearbook of the National Council of Teachers of Mathematics. Washington, D.C.: The Council, 1963.

Pace, Angela. "Arithmetic for the Fast Learner in English Schools." In *Enrichment Mathematics for the Grades,* pp. 193–94. Twenty-seventh Yearbook of the National Council of Teachers of Mathematics. Washington, D.C.: The Council, 1963.

Stephens, Lois. "An Adventure in Division." *Arithmetic Teacher* 15 (May 1968): 427–29.

Trimble, Harold C. "Problems as Means." *Mathematics Teacher* 59 (January 1966): 6–8.

Van Engen, Henry. "A Note on 'An Algebraic Treatment of Magic Squares.'" *Mathematics Teacher* 66 (December 1973): 747.

Watson, F. R. "Finding Out about Recurring Decimals." *Mathematics in School* 5 (March 1976): 29–30.

14

Teaching Computational Skills with a Calculator

Edward C. Beardslee

HOW many times have you seen or heard a statement similar to this: "The calculator is a threat to learning the basics—children will become so dependent on the machine that their minds will no longer function!" The calculator doesn't have to be a device that merely produces answers to computational exercises. It is exciting to discover *in the classroom* that the calculator *can* be used to help students learn mathematics. I have found that many students who had not been motivated to do mathematics using traditional methods are delighted to work with a calculator. And as a result, they develop needed mathematical skills.

The *Report of the Conference on Needed Research and Development on Hand-Held Calculators in School Mathematics* (NIE and NSF 1977) includes twenty-two recommendations concerning the calculator's role in mathematics instruction. Recommendations are listed under six categories (pp. 17–20):

- A. Development of an Information Base
- B. Curriculum Development for the Immediate Future
- C. Curriculum Development for the Long-Range Future
- D. Research and Evaluation
- E. Teacher Education
- F. Dissemination

Recommendation 6 under category B reads as follows (p. 18):

> Materials should be developed to exploit the calculator as a teaching tool at every point in the curriculum to test a variety of ideas and possibilities pending emergence of calculator-integrated curriculums.

This article provides ways to use the calculator as an integral part of mathematics instruction using your present curricular materials. However, the format of exercises must often be altered to require the pupil to think about what buttons need to be pushed rather than just pushing buttons. If each problem that the child takes to the calculator requires a decision regarding which buttons to push, then the device becomes a teaching tool. For example, standard exercises such as

$$\begin{array}{r} 36 \\ + 48 \end{array}, \qquad 87 - 62 = \square, \qquad 322 \div 23 = \square, \qquad \text{and} \qquad \begin{array}{r} 29 \\ \times 13 \end{array}$$

require only that the pupil push buttons to obtain the correct answer. But consider exercises that appear in this form:

$$\begin{array}{r} 3\,\square \\ + \square\,8 \\ \hline 8\ 4 \end{array}, \qquad 87 - \square = 62, \qquad \begin{array}{r} 2\ 9 \\ \times \square\ \square \\ \hline 3\ 7\ 7 \end{array}, \qquad \text{and} \qquad 34\,\overline{)\,6\,\square\,2}^{\,1\,\square}$$

Now the pupil needs to make decisions about which buttons to push in order to obtain the correct solution.

A first-grade student using the calculator to work exercises that require her to perform mental computations in order to get the final answers

Thus, in order to use the calculator effectively, we need to rethink the ways in which we are asking our students to solve exercises. To provide ways to get started, this article is devoted to activities and exercises that have been used with elementary school children. Topics are selected from different grade levels to illustrate how the calculator can be used to teach a broad range of mathematics topics.

Selected Calculator Activities

As we know, calculators differ. The activities cited here assume that a calculator with algebraic logic is used; that is, the machine requires that a mathematical sentence be entered into the machine in the way it is nor- mally taught (e.g., for 9 + 6 = 15, push 9, push $\boxed{+}$, push 6, push $\boxed{=}$, and the result appears). Some of these activities may need to be modified so they will work on your particular calculator. Try them out before using them with the class to see if any modification is necessary.

Counting

When children are learning to count and to recognize numerals, these images can be reinforced as the pupils push the buttons and see the numerals appear. Many calculators have a "constant" feature that en- ables the calculator to count: for example,

will result in a **3**. On some algebraic logic calculators, you can push

to reach **3**. Each additional push of the $\boxed{=}$ or $\boxed{+}$ will increase the display number by one. As soon as children are taught this trick, many will con- tinue to push $\boxed{=}$ to see how large a number they can obtain. One way to use this idea with a group of children is to instruct each child to push $\boxed{+}$ 1 $\boxed{=}$, then all say "one" together, push $\boxed{=}$ and all say "two", push $\boxed{=}$ and all say "three," and so on. Seeing the numbers as they count re- inforces their understanding of the counting process. This exercise can also suggest that addition is an operation rather than a relation. Since the machine displays the numbers, the children do not need to write them; thus, they can experience many numbers in a short period of time. Skip counting by twos, threes, fours, or by any number can be accomplished in the same way; for example, push ⊞ 3 ⊟ ⊟ ⊟ , and **9** is displayed. The following game helps to teach skip counting.

Game: **Guess my counter** (2 players)

Materials needed: 1 calculator, 1 score sheet, a clock with a second hand

Procedures: Players take turns. Without the second player seeing, the first player enters into the calculator an addition sentence using two one-digit numbers (e.g., 0 + 4 or 9 + 2). Then the first player presses ⊟ several times and hands the calculator to the second player. The second player continues to push ⊟ to try to determine the amount that is being added each time. If the second player correctly determines the counter within ten seconds, a point is scored. The winner is the first player to get ten points.

Reading numbers

Since a calculator presents a display of digits, it offers children a good opportunity to develop skills in reading numbers. For example, ask a student to enter any three-digit number and read it to you. Since the numbers can be changed rapidly, the student can practice many of them in a short period of time.

Read several numbers aloud. Direct each student to push ⊞ after each entry. After reading the last numeral in the list, check to see if the students obtained the correct sum. For example, read aloud: two hundred eight (please don't say "two-oh-eight"), seven hundred sixteen, one hundred thirty-five. Students who obtain 1059 when they push ⊟ have entered each of the numbers correctly.

Another activity that involves reading numerals uses word games. Word games involve entering numbers, then reading the display upside down. Often these word games are thought of as nonmathematical, but when they are used in the following manner, children have to make decisions and understand the terms in order to obtain the correct answers. I have found that many students have difficulty with these exercises because they make errors in entering the operations and numerals. If the exercise is presented orally, it helps to develop some important mathematical skills because children have to (1) listen carefully, (2) decide what operations need to be performed and in what order, and (3) understand the terminology used. So before discarding these exercises as useless, try presenting them orally to your students. If they do all the work correctly, the final answer will spell a word (or words) when the display is read upside down, and the student will enjoy a feeling of accomplishment. For example, read the following question aloud: "What did Mr. McGregor throw at Peter Rabbit to chase him out of the garden? To find out, add four to the product of twenty-seven and one hundred nine. Multiply by one thousand, subtract twenty-seven, and multiply by eighteen." The resulting words should be "his shoes."

Place value

Place value is a topic that often confuses children; hence, most teachers are seeking activities that reinforce it. Here are two games to help children practice place-value concepts.

Game: **Replace by zero** (2 players)
Materials needed: 2 calculators, score sheet

Procedures: Player A selects a three-digit number and directs player B to enter it in the calculator. Then player A directs player B to replace one of the digits by 0 without affecting any of the other digits. (For example, enter 273. Replace the 7 by 0. The student should subtract 70, leaving 203 in the display.) Players take turns, and those who do the operation correctly get one point. The first player to get ten points is the winner.

Game: **1000 wins** (2–4 players)
Materials needed: 2–4 calculators, score sheet, 1 die

Procedures: Player A enters three different digits in the calculator using only the digits 1, 2, 3, 4, 5, or 6 with no repeats—for example, 641. Player A then rolls the die. If the number on the die matches any of the three digits, player A scores the place value of that digit (e.g., if a 6 was rolled, player A would score 600; if a 4 was rolled, the score would be 40; if a 1 was rolled, the score would be 1). If the die number does not match any of the player's calculator digits, no score is obtained, and player B takes a turn. The player whose total score is closest to 1000 without going over is the winner. Players may decide to pass at any time prior to rolling a die.

Addition and subtraction

Many activities with the calculator are appropriate for both addition and subtraction. These sample activities include a variety of problems for (1) developing the concepts and (2) drilling the skills. As noted previously, the activities included are those that require a decision on the part of the pupil prior to pushing buttons. The thinking that the child must do to solve the problem is important if the calculator is to be used effectively. The following two activities simply present standard exercises in a slightly different format. Pupils must determine a starting strategy as well as decide when to add and subtract. As they work more exercises of this type, they should be encouraged to test and develop strategies for solving them.

1. Complete these exercises:

```
   4 5
   8 □              6 9 2
 + □ 9            − 1 □ □
 ───────         ─────────
   1 6 1            □ 2 3
```

2. Complete this addition table (fig. 14.1).

+			93
15		16	
	79		
	58		109

Fig. 14.1

The calculator can also be used to help drill the basic facts. One technique is to present a series of exercises orally, one at a time. Have the children work them on their calculators and call out each answer as soon as possible. As soon as you hear an answer, read the next problem in the series. If you are working on a particular fact, repeat it every third or fourth exercise. For example, if you wish to drill 9 + 7, try the following set of exercises: 4 + 5 (pause for an answer), 9 + 7 (pause), 3 + 5, 8 + 2, 9 + 7, 6 + 3, 7 + 4, 8 + 3, 9 + 7, and so on. This exercise takes only a few moments and adds some variety to drill. I have used this technique to teach that 17 × 15 = 255 (not usually considered a basic fact). Pupils are able to recall this fact several days later with no difficulty.

Another technique is to have the children enter an exercise, but before they push ⊟, call on one pupil to give the answer and then push ⊟ to check it. Let that child, if correct, give a problem and call on another pupil to answer. The teacher can begin by saying, "Enter nine plus seven. Before pushing the equal button, what do you think the answer will be, Jane?" When Jane answers, "Sixteen," everybody pushes ⊟ to check her. Then Jane chooses a problem and calls on another pupil to answer it.

Here is an example of another drill technique: "Enter twelve; what can you do to get five?" (Subtract 7.) "Start with eight; what do you do to get two?" (Subtract 6.) "Start with twenty-four; how do you get thirty-two?" (Add 8.) This type of exercise helps pupils develop problem-solving skills.

An activity that uses only particular numbers to generate the given sums or differences gives the child the opportunity to explore and solve problems by experimentation. As pupils develop strategies for completing these exercises, they will also be developing their problem-solving skills.

Use only the numbers 36, 8, 7, 11, and 12 to complete these sentences:

a. _____ + _____ = 15   d. _____ + _____ = 28
b. _____ + _____ = 20   e. _____ − _____ = 5
c. _____ + _____ = 47   f. _____ − _____ = 24

The following four games are just a few of the many ways of drilling addition and subtraction.

Game: **Three in a row** (2 players)

Materials needed: 1 calculator, 1 gameboard, and 10 markers (5 each of two colors)

Gameboard		Exercises

Gameboard

34	21	31
36	39	51
25	47	40

Exercises

13 + 27 = _____
15 + 19 = _____
7 + 18 = _____
16 + 23 = _____
22 + 9 = _____
37 + 14 = _____
8 + 13 = _____
24 + 12 = _____
13 + 34 = _____

Procedures: Players take turns selecting one of the exercises beside the gameboard. The player then solves the problem (using the calculator if necessary) and writes the answer in the space provided. The player then places a marker on the answer square. Play continues. The first person to get three markers in a row (across, down, or diagonally) is the winner. Note that the calculator can be used only after the exercise has been selected. (In watching pupils play this game, I observed that they were willing to follow the rules. They attempted to solve the exercise mentally before using the calculator.)

"Three in a row" is a version of ticktacktoe. After playing a version prepared by the teacher, pupils should be encouraged to construct their own versions and play them with each other.

Students playing "three in a row"

Game: **Top sum** (2 players)

Materials needed: 3 dice (two numbered 1 to 6 and the third with +, −, ×, +, −, × on the six faces), 2 calculators

Procedures: Players take turns tossing the three dice and mentally computing the result (e.g., if 2, 5, and + are tossed, compute 2 + 5; or if 3, 6, and × are tossed, then compute 3 × 6). This result is added to that player's display. The first player to obtain a display of 150 or higher is the winner.

"Top sum" requires students to use the calculator to keep track of the score while each player performs the calculations mentally.

"Beat the calculator" is a game in which the players challenge the calculator. Children find this activity stimulating and enjoyable.

Game: **Beat the calculator** (2 players)

Materials needed: 1 calculator, 1 score sheet, a clock with a second hand, and 45 cards with addition exercises (fifteen with two one-digit addends; fifteen with a one-digit addend and a two-digit addend; fifteen with two two-digit addends). The cards should have the answers on the back so that the answers can be checked easily.

Procedures: In fifteen seconds, player A shows as many as possible of the first set of fifteen cards (those with two one-digit addends) one at a time to player B, who gives the answer *without* using the calculator. At the end of the fifteen-second period, player A records how many were answered correctly. The task is repeated, but this time player B must use the calculator to obtain the answers. Now player A and player B reverse roles; player B shows the cards and records the number correctly answered in fifteen seconds. Each player's score is determined by adding the number answered correctly without the calculator to the number answered correctly with the calculator. The winner is the player with the higher score.

The players repeat the procedure using the second set of cards, then the third set of cards. Typically, pupils who know their addition facts will beat the calculator when using the first set of cards, will tie the calculator using the second set, and will lose to the calculator using the third set of cards.

The important aspect of this activity is that the players are engaged in mental drill, not whether the calculator wins or loses. This activity can also strengthen a pupil's ability to compute mentally with certain types of combinations.

Another popular and motivating game is a version of nim.

Game: **Variation of nim** (2 players)

Materials: 1 calculator

Procedures: The target number is 50. Players take turns adding any one-digit number to the display. The first player to obtain 50 in the display is the winner. Students should be encouraged to determine a winning strategy. Also, different target numbers should be used.

Multiplication and division

Many people feel that students should not use a calculator until they have a clear understanding of the concepts. However, the following examples suggest that concepts can be introduced using the calculator. Calculators with an addition constant and a subtraction constant are needed to do these examples.

Examples *(a)* and *(b)* use the calculator to develop the concept of multiplication as repeated addition.

a) 1. [+]3 [=][=] _____   2. [+]6 [=][=][=][=] _____
       How many 3s? _____   How many 6s? _____
       2 × 3 = _____   _____ × 6 = _____

    3. [+] 5   ?   **15**
       How many 5s? _____
       _____ × 5 = 15

b) Since multiplication is repeated addition, 27 × 16 = _____ can be solved by adding sixteen 27s together:
   [+]27[=][=][=] . . . (push [=] sixteen times).

Though most teachers introduce multiplication in this manner, usually the examples are simple; for instance, 3 × 4 is the same as 4 + 4 + 4. The calculator with an addition constant allows the student to illustrate multiplication through repeated addition by using larger numbers: 14 × 21 can be obtained by entering [+] 21 and pressing [=] fourteen times. This exercise takes only a few moments on the calculator.

   Examples (c) and (d) illustrate division as repeated subtraction.

c) 1. 15[−]5[=][=][=]0   2. 8[−]2[=][=][=][=]0
      Subtract how many 5s?_____   Subtract how many 2s?_____
      15 ÷ 5 = _____   _____ ÷ 2 = _____

   3. 18[−]3   ?   **0**
      Subtract how many 3s?_____
      _____ ÷ 3 = _____

d) Division is repeated subtraction: 798 ÷ 42 = _____. Use a calculator to repeatedly subtract 42 from 798. Count the number of subtractions (798[−]42 [=][=]. . .;count the number of times you push [=] until **0** appears in the display).

Again, exercises that are not normally illustrated this way are easily done on a calculator.

   Example (e) is similar to one of the addition and subtraction exercises.

e) Complete the multiplication table (fig. 14.2).

×	16		12
	192		
2		142	
		994	168

Fig. 14.2

The following game helps to drill multiplication facts.

Game: **Factor subtract** (2 players)

Materials needed: 1 calculator

Procedures: Enter 30, or some other number, in the calculator. Player A subtracts one of the factors of 30 other than 30 itself (possibilities are 1, 2, 3, 5, 6, 10, or 15). Player B then subtracts a factor of the number in the display. Players alternate subtracting a factor of the number in the display until **1** appears. The player who obtains **1** in the display is the winner.

Sample game: Start with 30
   Player A subtracts 15, leaving 15
   Player B subtracts 1, leaving 14
   Player A subtracts 7, leaving 7
   Player B subtracts 1, leaving 6
   Player A subtracts 3, leaving 3
   Player B subtracts 1, leaving 2
   Player A subtracts 1, leaving 1
   Player A is the winner.

The next exercise encourages pupils to investigate the multiplication algorithm.

Place the digits 2, 3, 6, and 8 in the blanks to obtain the highest product possible:

$$\begin{array}{r} \square\ \square \\ \times\ \square\ \square \\ \hline \end{array}$$

This exercise can also be used in a game format by using four cubes with numbers on the faces. Players arrange the cubes to obtain the largest possible product of two two-digit numbers. The product is their score. The exercise can be varied by using three digits and three boxes in this form:

$$\begin{array}{r} \square\ \square \\ \times\ \ \ \square \\ \hline \end{array}$$

Or five digits and five boxes in this form:

$$\begin{array}{r} \square\ \square\ \square \\ \times\ \ \ \square\ \square \\ \hline \end{array}$$

Pupils could also be asked to determine the smallest possible product rather than the largest.

The long-division algorithm is often difficult to teach. Use the calculator to work through each step of the algorithm to help develop this process. In this way pupils can work through several examples in a short period of time, thus developing a concept of the division process. See figure 14.3 for an example.

$$
\begin{array}{r}
138 \\
8 \\
30 \\
100 \\
\hline
47 \overline{)\,6486} \\
\end{array}
$$

47) 6486	
−4700	= 100 × 47
1786	= 6486 − 4700
−1410	= 30 × 47
376	= 1786 − 1410
− 376	= 8 × 47
0	= 376 − 376

Fig. 14.3

Problem solving

The calculator can be an extremely useful device for helping pupils develop problem-solving skills. When all pupils have the use of a calculator, they are essentially equal in terms of computational ability; that is, even though some have mastered the basic computational skills and others have not, the calculator helps to eliminate this discrepancy. With the calculator they can all compute with approximately the same speed and accuracy. Hence, children can be challenged to solve problems that they would not have been able to pursue otherwise simply because their inability to compute would have hindered their progress.

Another skill important to problem solving is the ability to estimate, approximate, and identify the reasonableness of results. This skill can also be developed using the calculator.

The topic of problem solving affords the teacher a tremendous opportunity to use the calculator. Problems that require tedious computations need not be ignored. Also, pupils can be encouraged to solve problems through exploration and trial and error.

Many of the exercises that have been presented under other topic areas could have been classified as problem-solving exercises. The examples listed here help illustrate a variety of exercise types that can be solved using the calculator.

a) Find the two-digit number that when added to its reverse is closest to 50 (e.g., 23 + 32 = 55); closest to 100; 150; 200; and so forth.

This problem is often solved by trial and error; however, as the target number is changed, say, 100 instead of 50, students begin to detect patterns and procedures for solving the problem. For example, if 13 is chosen as a solution for the target number of 100, then $13 + 31 = 44$, an answer that is not close to the target number. However, if 45 is chosen as a solution, then $45 + 54 = 99$. Also, $36 + 63 = 99$ and $72 + 27 = 99$. So, if the sum of the digits of a two-digit number is 9, then the sum of the two-digit number and its reverse (99) will be closest to the target number of 100.

b) In the following exercises, one wrong button was pushed to obtain the results. Can you find which one? (What button was pushed and what button should have been pushed? Remember that only one button was pushed wrong.)

$39 + 63 = 105$ (66 was added instead of 63)

$28 - 19 = 12$ (16 was subtracted rather than 19)

$8 \times 17 = 128$

$19 \times 23 = 247$

If pupils understand the arithmetic operations, then exercises of this type can easily be solved.

The next two problems are not typically solved using pencil and paper. Both, however, illustrate the many problems requiring extensive computation that can easily be solved using a calculator and therefore can easily be done in the classroom.

c) Use a clock with a second hand to determine the number of your heartbeats in one minute. How many beats an hour? A day? A year? How many in your life so far?

d) Count one count a second. How long will it take to count to one million? (eleven days, thirteen hours, forty-six minutes, and forty seconds)

The newspaper, an almanac, or the *Guinness Book of Records* can provide many interesting problems. For example:

e) According to the *Guinness Book of Records*, Tyrus Cobb scored 2244 runs in his career. How many miles did he travel to score these runs?

Problems of this type are easily generated, and the calculator provides a tool for helping to solve them. Pupils should be encouraged to write their own problems and share them with the class. In this way problem solving can become an enjoyable and exciting experience.

Here is another problem-solving example:

f) Try to write the numbers' from 1 to 100 using only the digits 1-7-7-6 in that order, using fundamental operations. For example:

$$1 + 7 + 7 - 6 = 9$$
$$1 + 7 + 7 + 6 = 21$$

A problem such as this will often excite students. A class could make this problem a class project, recording their solutions on the bulletin board.

Second- and third-grade students developing needed mathematical skills. Many students, who have not been motivated to work mathematics using traditional methods, are delighted to work with a calculator.

The last problem-solving example involves a series of computations that would be lengthy and tedious without a calculator:

g) If a sheet of paper could be folded in half twenty-six times, how many layers of paper would there be? (After one fold there would be two layers, after two folds, four layers, after three folds, eight layers, etc.) If the sheet of paper was 0.1 mm thick, how high would the stack be?

If a calculator has a constant multiplier feature, then the problem is easily solved by 2☒🟰🟰🟰. . ., pressing 🟰 twenty-five times. (The first time the 🟰 button was pushed, **4** was in the display, which represents the number of layers after two folds.)

Pattern investigation

The calculator provides the opportunity for investigating patterns that would be tedious and time-consuming with only paper and pencil. The examples that follow are merely a few of the numerous patterns that can be explored using a calculator. In working these problems, pupils need to write the answers at each step in order to identify the pattern as it develops. One method to do this is to prepare a work page giving the exercises and then leaving blanks for the answers.

a) $1 + 3 = ?$
    $1 + 3 + 5 = ?$
    $1 + 3 + 5 + 7 = ?$
    $1 + 3 + 5 + 7 + 9 = ? \ldots$

b) $1^2 = 1$
       $11^2 = 121$
       $111^2 = ?$
       $1111^2 = ? \ldots$

c) $\dfrac{1}{9} = ?, \dfrac{2}{9} = ?, \dfrac{3}{9} = ?, \ldots$

   $\dfrac{31}{99} = ?, \dfrac{32}{99} = ?, \dfrac{33}{99} = ?, \ldots$

   $\dfrac{812}{999} = ?, \dfrac{813}{999} = ?, \dfrac{814}{999} = ?, \ldots$

d) $1089 \times 1 = ?$
   $1089 \times 2 = ?$
   $\ldots$
   $1089 \times 9 = ?$

e) $1 \times 8 + 1 = \underline{\hspace{2cm}}$
      $12 \times 8 + 2 = \underline{\hspace{2cm}}$
     $123 \times 8 + 3 = \underline{\hspace{2cm}}$
    $1234 \times 8 + 4 = \underline{\hspace{2cm}}$
                $\ldots$
$12345678 \times 8 + 8 = \underline{\hspace{2cm}}$

f) $1^3 + 5^3 + 3^3 = \underline{\hphantom{mm}}$
   $8^4 + 2^4 + 0^4 + 8^4 = \underline{\hphantom{mm}}$
   $5^5 + 4^5 + 7^5 + 4^5 + 8^5 = \underline{\hphantom{mm}}$
   $5^6 + 4^6 + 8^6 + 8^6 + 3^6 + 4^6 = \underline{\hphantom{mm}}$

g) $1^3 + 2^3 = \underline{\hphantom{mm}}$
   $1^3 + 2^3 + 3^3 = \underline{\hphantom{mm}}$
   $1^3 + 2^3 + 3^3 + 4^3 = \underline{\hphantom{mm}}$

Decimals

Most of the examples given so far used integers. However, many of the exercises could easily be adapted to using decimals. For instance, in the counting examples, skip counting by 0.5, 0.25, or some other decimal number could be done. In the place-value activities, the numbers could be decimals instead of whole numbers. The addition, subtraction, multiplication, and division exercises could all be easily adapted to using decimals. Thus, a separate set of decimal-related activities is not included.

Conclusion

This article provides a variety of examples illustrating ways of using the calculator to help children learn mathematics. Topic areas were selected from the entire mathematics curriculum, grades 1 to 6. These illustrations are meant to encourage you to begin using the calculator in the classroom to help teach mathematics and to try out some new ideas and methods with your pupils. Also, as the mathematics curriculum is adjusted to meet future needs (e.g., teaching the metric system and placing more emphasis on consumer mathematics), you will be able to help your pupils acquire some of these skills through working with the calculator. Paper-and-pencil work will always be necessary to some extent, but the use of the calculator can help motivate youngsters to work with mathematics and provide a source of encouragement along the way so that success in learning mathematics skills can be reached. In many of the exercises outlined above, children who make errors or obtain wrong answers need only push the clear button and begin again. There is no written record of their errors; hence, errors do not become as much of an obstacle to learning as errors made in paper-and-pencil work.

This article has attempted to deal with only one part of the current mathematics curriculum, namely, integrating the calculator into the present curriculum. It is imperative that you, the classroom teacher, begin preparing for the future curriculum by being continually on the lookout for ways to use the calculator to help pupils develop needed skills.

BIBLIOGRAPHY

Beardslee, Edward C. "Calculators: Curse or Cure." *Washington Mathematics* 21 (January 1977): 1-4.

Beisse, Karen, Janet Brougher, and David Moursund. *Calculators in the Elementary School.* Portland, Oreg.: Oregon Council for Computer Education, 1976.

Immerzeel, George. *'77 Ideas for Using the Rockwell 18R in the Classroom.* Foxborough, Mass.: School Images, 1976.

National Institute of Education, and National Science Foundation. *Report of the Conference on Needed Research and Development on Hand-held Calculators in School Mathematics.* Washington, D.C.: NIE and NSF, 1977.

National Council of Teachers of Mathematics, Instructional Affairs Committee. "Minicalculators in Schools." *Arithmetic Teacher* 23 (January 1976): 72-74. Also available in *Mathematics Teacher* 69 (January 1976): 92-94.

Schlossberg, Edwin, and John Brockman. *The Pocket Calculator Game Book.* New York: William Morrow & Co., 1975.

Index